WITHDRAWN

B. F. HAMILTON LIBRARY
Franklin College
501 East Monroe Street
Franklin, IN 46131-2598

Practical Strategies for Experimenting

WILEY SERIES IN PROBABILITY AND STATISTICS

Established by WALTER A. SHEWHART and SAMUEL S. WILKS

Editors: *Vic Barnett, Noel A. C. Cressie, Nicholas I. Fisher,
Iain M. Johnstone, J. B. Kadane, David W. Scott,
Bernard W. Silverman, Adrian F. M. Smith, Jozef L. Teugels*
Editors Emeritus: *Ralph A. Bradley, J. Stuart Hunter, David G. Kendall*

A complete list of the titles in this series appears at the end of this volume

Practical Strategies for Experimenting

G. K. Robinson

CSIRO Mathematical and Information Sciences, Australia

JOHN WILEY & SONS, LTD

Chichester • New York • Weinheim • Brisbane • Singapore • Toronto

Copyright ©2000   John Wiley & Sons Ltd
                    Baffins Lane, Chichester,
                    West Sussex, PO19 1UD, England

                    National   01243 779777
                    International   (+44) 1243 779777

e-mail (for orders and customer service enquiries): cs-books@wiley.co.uk

Visit our Home Page on http://www.wiley.co.uk or http://www.wiley.com

All Rights Reserved. No part of this publication may be reproduced, stored in a retrieval system, or transmitted, in any form or by any means, electronic, mechanical, photocopying, recording, scanning or otherwise, except under the terms of the Copyright, Designs and Patents Act 1988 or under the terms of a licence issued by the Copyright Licensing Agency, 90 Tottenham Court Road, London W1P 9HE, UK, without the permission in writing of the Publisher and the copyright owner, with the exception of any material supplied specifically for the purpose of being entered and executed on a computer system, for the exclusive use by the purchaser of the publication.

Designations used by companies to distinguish their products are often claimed as trademarks. In all instances where John Wiley & Sons is aware of a claim, the product names appear in initial capital or all capital letters. Readers, however, should contact the appropriate companies for more complete information regarding trademarks and registration.

Other Wiley Editorial Offices

John Wiley & Sons, Inc., 605 Third Avenue,
New York, NY 10158-0012, USA

Wiley-VCH Verlag GmbH
Pappelallee 3, D-69469 Weinheim, Germany

Jacaranda Wiley Ltd, 33 Park Road, Milton,
Queensland 4064, Australia

John Wiley & Sons (Asia) Pte Ltd, 2 Clementi Loop #02-01,
Jin Xing Distripark, Singapore 129809

John Wiley & Sons (Canada) Ltd, 22 Worcester Road,
Rexdale, Ontario, M9W 1L1, Canada

Library of Congress Cataloging-in-Publication Data

Robinson, G. K.
    Practical strategies for experimenting / G. K. Robinson.
      p. cm – (Wiley series in probability and statistics)
    Includes bibliographical references and index
    ISBN 0-471-49055-5 (alk. paper)
    1. Experimental design. I. Title II. Series
    QA279 .R64 2000
    001.4'34—dc21   00-027336

British Library Cataloguing in Publication Data

A catalogue record for this book is available from the British Library

ISBN 0-471-49055-5

Produced from files supplied by the author and processed by SunRise Setting, Torquay
Printed and bound in Great Britain by Antony Rowe, Chippenham.
This book is printed on acid-free paper responsibly manufactured from sustainable forestry in which at least two trees are planted for each one used for paper production.

Contents

Preface vii

1 Introduction 1
   1.1 Ideas Behind the Structure of the Book 2

2 Clarify the Objective 9
   2.1 Consider who to Consult 10
   2.2 Ask Questions 14
   2.3 Statistical Models and Types of Experimental Objectives .. 22
   2.4 Taguchi'S Quality Engineering Ideas 25

3 Summarize Beliefs and Uncertainties 29
   3.1 Decide whether you Need an Experiment 29
   3.2 Consider Relevant Science and Technology 32
   3.3 Sampling of Particulate Materials 34
   3.4 Consider Information from Existing Data 39
   3.5 List Response Variables and Factors 48
   3.6 Consider Possible Transformations of both Response Variables and Factors 51
   3.7 Document your Summary of Beliefs and Uncertainties 58

4 Decide on a Strategy 61
   4.1 Main Effects and Interactions and their Ranges of Validity . 67
   4.2 Beware of One-Factor-at-a-Time Experimentation 71
   4.3 Strategies for Studying Sampling and Testing Errors 75
   4.4 Main Effects Experiments 79
   4.5 Case Study: an Experiment on a Wave Soldering Machine . 81
   4.6 Response Surface Methodology 87
   4.7 Case Study: a Simulation Game 94

5 Plan a Single Experiment 113
   5.1 Decide which Response Variables to Measure 113
   5.2 Search for Efficiencies 116
   5.3 Decide When, Where and With What Materials 124
   5.4 Take Precautions to Avoid or Allow for Biases 127
   5.5 Cope with other Sources of Variation 133

    5.6   Case Study: Surface Treatment of Tool Steel 140

6 Design the Experiment 143
    6.1   The Concepts of Confounding and Orthogonality 143
    6.2   Issues Affecting the Number of Runs 148
    6.3   Some Types of Experimental Designs 155
    6.4   Decide which Factors to Include and at what Levels 175
    6.5   Choosing a Design 179
    6.6   Case Study: Ina Tile Experiments 181

7 Collect the Data 189
    7.1   Set up a Data-Recording System 189
    7.2   Using Spreadsheets for Data Recording 191
    7.3   Try to Anticipate Possible Problems 194
    7.4   Perform Runs and Record what Happens 196

8 Update Beliefs and Uncertainties 199
    8.1   Cross-examining the Data 200
    8.2   Focus again on your Purpose 203
    8.3   Describing Trends 204
    8.4   Describing Variability 215
    8.5   Checking Assumptions 219
    8.6   Describing the Balance between Confidence and Uncertainty 230
    8.7   Using Computers for Statistical Analysis 234
    8.8   When to Seek Help with Statistical Analysis 235

9 Revisit the Objective 239
    9.1   Reconsider Objective and Planned Strategy after each Experiment . . 239
    9.2   Communicate Results 241
    9.3   Help Make Changes Happen 246
    9.4   Contribute to Improving other Things 247

Glossary 251

References 259

Index 263

Preface

My purpose in writing this book has been to help people to plan programmes of experiments more effectively. I have not ignored technical statistical topics such as design and analysis of experiments, but my primary focus is not on technical issues. Although my primary training is as a statistician, I believe that the non-statistical aspects of planning and conducting experiments are more important than the formal design and analysis. These aspects include answering questions that actually matter, deciding what to measure, considering what might go wrong and taking some personal responsibility for ensuring that changes are actually implemented.

The people who I have principally had in mind when deciding what material to include in the book are people whose experimental objectives I have thought about at various times over the years. I might have worked with them as a statistical consultant, as a colleague, or because they were attending a training course that I was helping to conduct.

These people often have tertiary qualifications in scientific or engineering disciplines. They are generally satisfied that they have enough mathematical skills for the tasks that they need to perform. They have usually attended courses in statistics, but not gained the skills that they now feel they need to design and conduct experiments.

My book differs from most other books about design of experiments and analysis of experimental data in that it concentrates on questions for which there are no uniquely correct answers. (In my opinion, too much of mathematical and statistical training concentrates on problems which have right answers, thereby ignoring the relationship between reality and theory which I consider to be the most useful and interesting part of these subjects.) This focus has influenced both the content and the style of the book.

- I often present opinions or value judgements, in order to provide guidance when there is not a *best* way to proceed.
- I often suggest questions that might be worth considering, rather than providing answers.
- I have included a lot of examples of situations which illustrate the issues being discussed. Unlike the examples in many other books, very few of these are numerical examples for readers to test their ability to perform calculations.
- Generally, I have not included statistical theory or computational methodology. With the exception that I have included some theoretical material about sampling of particulate materials. I included this because I believe that sampling is more often a constraint to the precision of an experiment than many people realize and the theory is somewhat inaccessible.

- I have arranged topics in an order in which they might be considered when planning and conducting a programme of experiments, rather than in order from mathematically simple to mathematically complex. A good way to use this book would be to read it during the planning of a substantial experimental programme, checking that important issues are not being overlooked.
- The book is not intended to be used as an introductory text. It assumes that readers have already encountered most of the concepts discussed, perhaps because they have already read an introductory book on the design and analysis of experiments, and have previously conducted experiments themselves.

From my point of view, this book has its origins in discussions with people about their research. For many of these discussions, my professional role was that of statistical consultant.

The second stage of the book's evolution was during the time I presented many training courses on experimentation, especially as a result of the feed-back I received from co-presenters and from people attending the courses. The development of my ideas was strongly influenced by my co-presenters, particularly Teresa Dickinson, but also Allan Adolphson, Paul Livingstone, Aloke Phatak and Karen Burrows. My ideas were also influenced by Alan Long who invited me to watch training courses which he was conducting.

The third stage in the evolution of the book was the conversion of my training course materials into a book, omitting or replacing material which was effective in the interactive environment of a training course but would not work particularly well in print. I would like to thank Sue Chambers, Sarah Darby, Tom Fearn, Bronwyn Harch, Ross Hughes, Richard Jarrett, John Maindonald, Warren Muller, Nam-Ky Nguyen, Ray Reynoldson, Albert Trajstman and Alan Veevers, who contributed in various ways to examples given in the book. I would also like to thank the many people who commented on drafts, especially Neil Fothergill and John Maindonald.

It is now time for me to stop revising, even though the ideas presented are continuing to evolve.

GEOFF ROBINSON
Melbourne, Australia

1
Introduction

Many books have been written about theoretical design and analysis of experiments. This book has a wider scope but less mathematical depth. It gives anecdotes and check-lists of suggestions about experimenting. It concentrates on examples and value judgements, with little discussion of computational techniques or experimental designs.

It is structured as a generic strategy for experimenting. I hope that readers will use the book as a guide during their experimentation, to check that they haven't forgotten to think about any important issues.

This book is intended primarily for people who will conduct experiments. It is intended to help them to plan experimental strategies: this includes designing and analysing simple experiments, and summarizing the results. The book should also be of some interest to the supervisors of experimenters, to statisticians, to the teachers and trainers of such people and to people interested in using information from experiments.

It presents an approach to experimentation at the strategic level. It describes broadly how to proceed and what questions to think about. It provides check-lists of things that might otherwise be forgotten. It does not provide many technical details.

One feature of this book that I hope will appeal to most readers is that it provides many examples to illustrate the ideas discussed. Many of them are based on real experiences, often simplified so that they better illustrate a particular point.

A second feature of this book is that it emphasizes the qualitative aspects of experimenting. You need to remember to do things like the following.

- Consult people about the nature of the real problem to be solved and about what they already know. It can be very disheartening to find that results of a complex experiment are not relevant or are already known.
- Ask good questions and formulate useful hypotheses. Delaying speculation and questioning until late in an experimental programme can be a very inefficient way to proceed.
- Recognize the advantages of experiments over observational studies. Deliberately vary controllable factors in order to get information and vary many factors in a balanced way so that their effects can be estimated separately from one another.
- Plan to conduct an experimental programme in stages in order to minimize the impact of unforeseen problems. A small, preliminary trial to check experimental procedures may turn out to have been a minor waste of resources, but failure to conduct such a small, preliminary trial may lead to such a major waste of resources that the experimental programme cannot be completed.
- Choose the conditions to be investigated thoughtfully. Sometimes radical choices are appropriate so that experimental programmes can either quickly provide new

directions or be terminated. Sometimes caution must reign. Some experiments should be done with ultra-pure water, while other experiments should be done with the dirtiest water likely to be subsequently encountered.
- Recognize that all trials should be regarded as experiments and planned carefully. An experiment is any deliberate interference with a process in order to gain information about that process. It does not have to involve test-tubes and white coats. Some experimenters spend a lot of effort trying things out before doing what they regard as a formal experiment. The 'trying things out' phase needs to be considered as experimentation, and to be just as important as the formal experiment.
- Report experiments with emphasis on what might be done differently in the future. Some experimenters fail to acknowledge any responsibility for ensuring that their results are translated into action.

The ideas in this book have evolved from my personal experiences of statistical consulting. The form of the book has evolved from a number of talks and training courses that I have given over the last 15 years. My primary motivation for presenting those talks and training courses and for writing this book is that I have seen many experiments which were wasteful of resources, and which often had little chance of producing information adequate for the intended purpose. They were disasters which might reasonably have been foreseen. Some of them are mentioned later in the book.

I believe that many experimenters need guidance on qualitative aspects of experimentation and with experimental strategies. I don't know all the answers. However, I believe that I am aware of the most important questions and aim to help readers of my book to learn to experiment more effectively. I do not mean that you should learn complex techniques or be able to pass some examination. I mean that you should remember to stop and think about important issues and that your disasters should be smaller than they might otherwise have been.

This book is intended for readers who already have some experience of scientific or industrial experimentation and who also have some knowledge about techniques for the statistical analysis of data. I would encourage you to read the book at the same time as you plan and conduct a programme of experimentation, considering the implications of the questions raised by the book as applied to that programme of experimentation.

The book is not organized like a mathematical textbook which defines new concepts, develops them and then illustrates them. Instead, topics and themes are introduced and discussed in an order in which they might be relevant to a programme of experimentation. One difficulty with this approach is that it is often necessary to plan something long before it is executed, so discussions about planning may require familiarity with concepts which are discussed in more detail later in the book. The index and the glossary should help to alleviate this difficulty.

1.1 IDEAS BEHIND THE STRUCTURE OF THE BOOK

This book is structured following a possible order for the tasks involved in planning and executing a programme of experimentation. Three diagrams are discussed in this section to explain the basis for the structure chosen.

1.1.1 A Diagram of how Problem-solving Proceeds

One way of thinking about solving problems is illustrated by Figure 1. In this model, problem-solving has three stages:

1. **extraction** – a theoretical problem for which a solution is thought feasible is extracted from its multi-stranded real context;
2. **solution** – a solution is found for the theoretical problem as extracted;
3. **application** – the solution to the extracted problem is applied back to reality.

Good extraction of problems involves understanding customers' situations, priorities and processes. It requires knowing what problems are likely to be resolvable and it involves anticipating how a solution might be applied or delivered. Extracting strategic research problems which are likely to have many potential applications is generally harder than extracting tactical problems, because it requires a broader understanding of existing situations, priorities and processes.

Good application needs to include the reporting of uncertainties and thoughts about where to go to next as well as reporting solutions. It may involve a great effort to get changes to actually happen. Researchers don't have to do much of this, but they should not completely divorce themselves from knowing whether or not it happens.

Some people have used similar models. For instance, Senn (1998) described a chain of reasoning which proceeds from a 'real problem' via an 'operational problem' and a 'solution' to an 'application'. Polya (1945) presented a similar approach to solving mathematical problems.

Figure 1 is intended to illustrate that extraction and application usually involve loss, in the sense that the amount of progress on a real problem is seldom as great as would be suggested by looking only at the apparent progress made by the theoretical solution to the extracted problem. The amount of such loss should be considered. For instance, theoretical solutions to small problems may be of no value at all.

The ideas implicit in this diagram lead me to feel that some people who regard themselves as 'mathematicians' and 'statisticians' take too narrow a view of the scope of their subjects. The content of courses, the content of academic papers and the community perception of the professional life of mathematically oriented people all emphasize the solving of theoretical, well-posed problems. I believe that this narrowness has restricted the usefulness of such people.

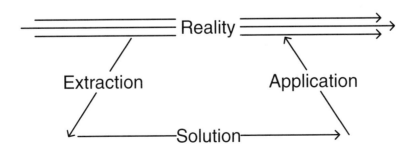

Figure 1. Diagram of how problem-solving proceeds

I am sympathetic to the philosophy of 'greater statistics' rather than 'lesser statistics' as presented by Chambers (1993). Many other people have expressed similar views. I believe that extraction should be regarded as the most important part of problem solving. It is the key to being able to solve problems expeditiously and also the key to effective application of solutions. This book gives little emphasis to the solving of theoretical problems, on the grounds that there are many other books available which do an excellent job of explaining the theories of experimental design and statistical analysis of data. Some other books about experimental design and statistical analysis which are written from standpoints similar to mine are Cox (1958), Mead (1988) and Pearce (1976).

1.1.2 A Diagram of how Science Proceeds

Box (1976) has made some useful comments about aspects of scientific method that I would like to build upon. The first of them emphasizes iteration between theory and practice. He argues that science is a means of learning which is more than theoretical speculation and is more than the undirected accumulation of practical facts. It makes progress by the alternation of two phases, deduction and induction.

- **Deduction** starts with some aggregation of theories, models, conjectures, hypotheses and ideas. It makes predictions and suggests ways of observing the world and collecting data.
- **Induction** starts with some observations, data, facts and knowledge about practices. It chooses between theories and models based on that information. It may also build on previous theories, models and ideas.

A crucial step in induction is judging whether discrepancies between predictions and data are such that previously accepted theories or models are no longer acceptable. Kuhn's (1970) ideas are well worth considering; he reviewed a large amount of historical evidence about major scientific advances and conflicts between rival theories (and based his models and conceptualizations on how science has proceeded, rather than starting with an introspective opinion as to how science ought to proceed). He considered that it is useful to distinguish between periods of 'normal science' during which problems are solved without challenging the generally accepted, coherent collections of theories (which he refers to as 'paradigms') and 'periods of crisis' and 'revolutions' when substantial quantities of old theory are discarded and replaced by new theories (which he describes as the 'emergence of a new paradigm').

Figure 2 is an attempt to illustrate the ideas of Box and Kuhn in a single picture. It shows a period of 'normal science' during which the deductive steps make predictions which vary only slightly according to the various hypotheses, ideas and conjectures which are being considered. The inductive steps take observations, data and facts, and draw conclusions about which of the hypotheses, ideas and conjectures are likely to be accurate, true or useful approximations. Results of either experiments or observational studies are not usually replicated during periods of 'normal science', because there is little conflict between the steadily evolving ideas of the various researchers in the field of study.

A 'period of crisis' happens when the generally accepted models, theories and paradigms do not seem to be able to predict some aspects of what has been

INTRODUCTION

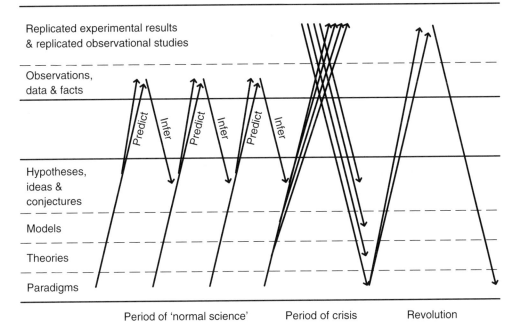

Figure 2. A diagram of the irregular progress of science

observed. The data is usually doubted before theories and paradigms are queried, so during a period of crisis researchers often replicate both experimental results and observational studies. During such a period of crisis, deductive predictions might be made considering alternative models, theories or paradigms and the predictions will generally differ. The inductive steps made during periods of crisis are not accepted by all researchers in the field of study. They might not all accept that all of the data has been adequately replicated. Some researchers make inferences about minor hypotheses or conjectures, without casting any doubts on their strongly held beliefs about models, theories and paradigms. Other researchers infer that models, theories or paradigms need to be overthrown. Alternative deductive and inductive steps are shown in Figure 2 as arrows.

What Kuhn calls a revolution happens when the deductions from various paradigms are different, and induction from reproducible data indicates that a change from the previously accepted paradigm is necessary. Less dramatic forms of revolution can also occur, when less basic models and theories are overthrown.

1.1.3 A Diagram of an Experimentation Checklist

There are many steps which can be useful in the planning, conduct and analysis of experiments. I have presented check-lists of between 22 and 24 steps in training courses for experimenters, and have used similar check-lists in statistical consulting for several years. Comments from many people have been useful in the refining of these checklists. I would particularly like to acknowledge the contributions of Teresa Dickinson, with whom I have presented many training courses. One such check-list is the following.

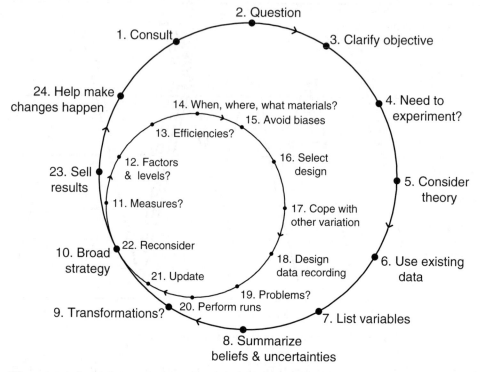

Figure 3. A diagram of a checklist for experimenting as an inner cycle within an outer cycle

1. Consider who to consult.
2. Ask questions.
3. Clarify the objective.
4. Decide whether you need to experiment.
5. Consider relevant science and technology.
6. Consider information from existing data.
7. List response variables and factors.
8. Summarize beliefs and uncertainties.
9. Consider possible transformations of both response variables and factors.
10. Decide on broad experimental strategy.
11. Decide which response variables to measure.
12. Decide which factors to include and at what levels.
13. Search for efficiencies.
14. Decide when, where and with what materials.
15. Take precautions to avoid biases.
16. Select an experimental design.
17. Cope with other sources of variation.
18. Design a data-recording system.
19. Try to anticipate possible problems.
20. Perform runs and record data.

INTRODUCTION

21. Update beliefs and uncertainties.
22. Reconsider objective and planned strategy.
23. Summarize and communicate results.
24. Help make changes happen.

Similar check-lists have been given by Coleman and Montgomery (1993), Daniel (1976), Hahn (1984), Hunter (1988), Jeffers (1978), Knowlton and Keppinger (1993) and Whitney and Young (1989). Such check-lists should not be regarded as uniquely correct or exhaustive. They are not equally applicable to all problems. And sometimes it may be best to follow a standard set of steps in a non-standard order. However, they do provide a basis for thinking about strategies for experimenting.

The check-list of steps may be regarded as an version of the Plan-Do-Check-Act cycle modified to have an inner cycle within an outer cycle. The outer circle is concerned with the overall experimental programme and the inner circle is concerned with the conduct of a single experiment. This is shown in Figure 3.

Many courses on the statistical part of conducting experiments concentrate on the experimental design and data analysis. These are shown as 'Select design' and 'Update' in the inner circle.

My experience is that successful experimentation requires that both circles function well. It is particularly important that you do not lose sight of the overall objective of the experimental programme (the outer circle) while coping with the large number of tasks which need to be done as parts of the work of conducting experiments (the inner circle).

In this book, the steps on the check-list have been amalgamated into chapters, so that each chapter corresponds to a milestone in a strategy for experimenting.

2

Clarify the Objective

Clarifying the objective is a crucial part of any project. Tukey (1962, pages 13–14) wrote:

> Far better an approximate answer to the right question, which is often vague, than an exact answer to the wrong question, which can always be made precise.

Clarifying the objective usually involves consulting a wide range of people, including those affected by the problem to be solved, the people who will conduct the trials and the people who might introduce modifications. One reason for consulting them is to check that that your impressions about the real objective are accurate. It is particularly important to identify and consult the problem owner. This might be a single person, a group of people, or a committee. The problem owner is the gatekeeper who decides about funding or who decides whether changes in procedures will be adopted.

Clarifying the objective also involves asking a wide range of questions as steps towards preparing a clear statement of your objective. Such a statement should do all of the following.

1. Present choices for doing something differently in the future.
2. Specify the range of situations to which your conclusions could be applied.
3. Discuss how outcomes might be evaluated.
4. Describe what precision and complexity are acceptable.

In order to present the material in this book in a tidy fashion, I have included 'clarifying the objective' as a milestone in conducting a programme of experiments. In practice, the objective is often found not to be completely clear after you have moved on to other phases of experimental planning. Sometimes the objective of a programme of experiments is quite clear very early. For other programmes, the objective seems to change continually. You must be prepared to reconsider the objective at a later stage if this seems to be necessary.

I find it helpful to distinguish between three types of experimental objective, as discussed in more detail in Section 2.3 on page 22.

- Sometimes your objective is to get a relatively simple model; perhaps just to find out which factors have any effect, perhaps to fit a simple model for the mean or the trend. I refer to this as 'main effects experimentation'.

- Your objective may require finding a model for the trend in a response which is more accurate than previous models, accuracy being more important than simplicity. I refer to this as 'response surface experimentation'.
- Perhaps your objective is more about describing variation than about describing a trend. I refer to this as 'estimating the precision of sampling and testing'. (A more abstract description might be more accurate, because some such experiments are not concerned with sampling or testing, but this terminology is easy to understand.)

The boundaries between these types of objectives are not precisely defined, and many real experimental objectives include aspects of all three of these types. Despite these difficulties, I believe that many people will find the categorization to be helpful.

2.1 CONSIDER WHO TO CONSULT

Consider who might be able to help you with the project. Construct a list of possible helpers. Include people with a variety of views. Include people who between them have all of the skills required. Include representatives of groups who might be expected to change their opinions or their actions as a result of the knowledge obtained.

Your selection of people will be based on an initial specification of the problem and the objective. This specification may change – so be willing to consult other people as more opinions and data arrive.

2.1.1 Starting

The origins of a programme of research are usually difficult to state precisely. Perhaps a process is performing unsatisfactorily. Maybe some information is needed in order to write a specification. Perhaps an individual thinks of a radical idea that seems to be worth trying. Information may be needed to help make decisions about proposed new equipment or procedures.

Sometimes, I record half-baked ideas for possible future research in a folder labelled 'Maybes'. Nothing happens to most of these ideas, they are discarded after a few years. But some of them get supported by someone else's ideas, by my interest in a related problem, or by a development in methodology that makes one of my 'Maybes' seem more achievable or more worthwhile. Then I am likely to consider them again.

Other people have different ways of handling half-baked ideas. Many people make no record of their partly formed ideas about what or how something might be achieved.

I would encourage readers to consider whether it would be possible to improve their personal procedures and their organization's procedures for recording and developing half-baked ideas. However, this is outside my expertise and outside the focus of this book.

Often, the first clear evidence that a programme of research might be starting is when someone argues that the likely benefits of investigating a problem and conducting experiments exceed the likely costs. This is often a difficult argument to make because likely costs are generally easier to quantify than likely benefits. I suggest that this argument is more likely to succeed if it is delayed until after other people have been enlisted to discuss the possible research programme. This is a major reason why my experimentation checklist starts with considering who to consult.

CLARIFY THE OBJECTIVE 11

Considering who might help is the first substantial step between a nebulous state of thinking that something might happen sometime and a state of deliberate intent to make progress. You will feel emotionally that you have started once you discuss a possible project with other people. You have made tangible progress merely to have convinced other people that your problem is worth thinking about.

2.1.2 Identify the Problem Owner

There may be a single person, a group of people or a committee who could be regarded as the problem owner. A well-defined problem owner should have authority to ensure that experimental runs proceed and also have authority to ensure that if changes are recommended then they will be implemented. If such a problem owner can be identified then that person (or those people) must be consulted.

In many situations there is not a well-defined problem owner. Permission for experimental runs to proceed might need to be sought from several parties.

- Funding might come from more than one source. This can be particularly complicated when some funding comes from a government.
- Equipment availability might need to be checked.
- Access to fixed plant, agricultural land, water, or similar resources might need to be checked with other possible users of those resources.
- Environmental, safety or ethical issues may need to be cleared with parties which are separate from the normal management or administration.

Similarly, there may be many parties whose cooperation must be sought in order to implement possible changes.

- Funding for capital works might be required.
- Employee or union cooperation may be required if there might be changes in work practices.
- Cooperation of suppliers may be required if there could be changes to input specifications.
- Cooperation of customers may be required if the characteristics of outputs might change.

2.1.3 Why Consult?

There is often a trade-off between time spent solving a problem and time spent implementing a solution. By increasing the participation of interested and potentially affected people in the problem-solving process it is generally possible to save time in implementation.

It may be possible to solve a problem by yourself. Perhaps, other people will have relevant information which you will need to obtain. Also, other people may need to be persuaded that your solution is correct, either before or after it is implemented.

Participation in problem solving and decision making generally encourages commitment to the agreed solutions/decisions. The greatest possible sense of commitment will be achieved if the people responsible for implementing a solution feel that they own the problem and the solution. For this to happen, people outside this group should be seen to be involved primarily as facilitators in the decision-making or problem-solving process, providing technical skills, judgement, experience, information and advice when these are requested.

In all circumstances, it is advisable not to put anyone into a position where it is in their interests to see the possible changes fail.

2.1.4 Compiling a List of People

The set of people to be consulted about a project should not be called a 'project team'. That term might be used to describe a group of people who work together and who all spend a substantial proportion of their time on the project. Some of the people to be consulted may never meet, may spend little time on the project, and may even be unaware of its existence.

The list of people is likely to change over the life of the project. It should include:

- people with sufficient skills to plan, conduct and analyse experiments;
- people whose prior opinions cover the range of views held by people whom you hope will be influenced by the conclusions to be reached eventually;
- people with enough influence to make experiments happen and to sell the conclusions drawn.

Being on the list of people to be consulted does not mean contributing to the effort of conducting the experimental runs. It does mean contributing opinions, guidance or influence. It does mean being prepared to discuss the available options, any constraints, and the possible negative effects of things that might go wrong.

Management assistance may need to be sought if some of the people to be consulted are in other organizations.

2.1.5 Example: Preservation of Grapes

This is an example of how an experimental programme can fail badly due to inadequate consultation.

An experiment started out with a desire to compare a standard method of preserving table grapes to an alternative method. The grapes were sultanas to be sold as fresh grapes. The standard preservation method was to store grapes in a coolstore with some sulphur dioxide in the atmosphere. The proposed alternative was to place grapes in plastic bags with pellets which slowly release sulphur dioxide. There were to be small holes in the plastic bags to allow slow diffusion of gases into and out of the bags.

The experimenter's supervisor considered that a scientific experiment like this needed to have a 'control', but wasn't sure what was meant by a control in this situation. The experimenter wasn't sure either. Perhaps using a control meant using no sulphur dioxide. Perhaps it meant using no plastic bags; maybe it meant using plastic bags with no holes? (I think that an experiment like this in which a proposed alternative procedure is being compared to a standard one does not require a control, but that is not my current focus.)

What treatments were actually used in the experiment? First there were the two that were really necessary:

CLARIFY THE OBJECTIVE

- sulphur dioxide fumigation;
- putting pellets which release sulphur dioxide with the grapes, inside plastic bags with small holes.

Several other treatments were also used:

- sulphur dioxide fumigation, bags with no holes;
- sulphur dioxide fumigation, bags with holes;
- sulphur-dioxide-releasing pellets, no bags;
- sulphur-dioxide-releasing pellets, bags with no holes;
- no sulphur dioxide, no bags;
- no sulphur dioxide, bags with no holes;
- no sulphur dioxide, bags with holes.

The experiment was intended to last for several months, but most of the grapes from these 'other' treatments went rotten within two months. This was a problem from the coolstore owner's point of view. He had donated space for the experiment. He was charging commercial rates for the adjacent areas, yet the customers renting the adjacent areas thought that having their grapes next to the rotten experimental grapes was unsatisfactory. Soon, it was agreed to assume that rotten grapes could be predicted to remain rotten, and hence to score zero for marketability for the remainder of the experimental period, so they could be thrown out.

My reason for discussing this example is not to discuss the use of controls, but to discuss the need to consult appropriate people. The only people consulted about this experiment were:

- the experimenter (for his expertise), and;
- the experimenter's supervisor (apparently for his views).

Who should have been consulted?

- **For their skills:** the experimenter and his supervisor.
- **For their views:** the coolstore owner, a grape grower, a wholesaler, a retailer and the sulphur dioxide pellet salesman.
- **For their influence:** the Director of the research institute and an industry representative.

If even a few of these people had been consulted then the experiment might have been different. For instance, the treatment which attempted to preserve grapes by putting them in plastic bags with no holes and having sulphur dioxide outside these bags seems very silly and would have been difficult to defend in discussions with any of these people. Other changes might also have been suggested. A retailer might have suggested that grapes would be left in the bags with the pellets after they were taken out of the coolstore, so that it would be sensible to measure grape quality at a different time.

This example is continued on page 21, in order to illustrate how asking questions can help to clarify an objective.

2.1.6 Example: Fire Ecology Experiment

A long-term fire ecology experiment conducted in northern Australia has been described in Andersen *et al.* (1998). It was motivated by the belief that much information relevant to fire management in northern Australia has been derived from observations over inappropriately small spatial and temporal scales. Warren Muller, a colleague who was involved in this experiment, discussed this example with me.

The following four treatments were used for a five-year period, after two years in which fires had been excluded from the area.

1. 'Early burn': fires were lit early in the dry season, like the commonest current management regime.
2. 'Progressive burn': areas were burnt progressively as they dried out during the dry season, approximating the traditional burning practices of Australian aboriginals.
3. 'Late burn': fires were lit late in the dry season, approximating unmanaged 'wildfires'. Such fires tend to burn at high temperatures and are very difficult to control.
4. 'Unburnt': no fires were lit as part of the management regime. However, the endeavour to have no fires was expected to be upset by accidental conflagrations.

One region treated as 'Late burn' and one treated as 'Unburnt' were burnt by unplanned fires during the five years of the experiment.

Thirteen regions of between 15 and 20 km^2 were designated as experimental plots. The boundaries between the regions were low ridges. Four regions were assigned to the 'unburnt' treatment and three were assigned to each of the other treatments.

The assignment of regions to treatments was only partly randomized. Andersen *et al.* (1998) says that the randomization was 'subject to topographic and security constraints'. The two regions adjacent to a highway may have been constrained to being 'early burn' or 'progressive burn' plots, so that tourists travelling along the highway would not be critical of forest management practices, and because the risk of accidental fires is greatest adjacent to the highway. Late burns need to be stoppable, and the selection of 'late burn' plots may have considered this constraint.

Six core research projects focused on nutrients and atmospheric chemistry, temporary streams, vegetation, insects, small mammals, and vertebrate predators. They had different requirements for sampling and different logistic constraints. Yet at the end of the experiment they need to be telling a coherent story.

For an experiment like this, with many groups of people having different interests and facing different practical difficulties, consultation is a critical step in planning. If important decisions and compromises had been made without consulting interested parties then the experiment would not have been as successful.

2.2 ASK QUESTIONS

During the early stage of planning experiments you should ask questions to check that everybody understands:

- the context of the proposed experimentation;
- the outcome sought;

CLARIFY THE OBJECTIVE

- its practical consequences, and;
- environmental, safety and ethical issues.

2.2.1 What is the Organizational Context of the Proposed Experimentation?

In order to justify the resources required for conducting experiments you will need to be able to explain the potential benefits. Barton (1997) advocated using what are called Goal Hierarchy Plots for thinking about the goals of experimental programmes and how they fit within organizations' broader goals. These plots display high-level business goals at the top and display several levels of subgoals which are required in order to achieve the higher-level goals above them. Such plots can be useful for relating the objectives of an experimental programme to the objectives of the various people who might be consulted about the experiments.

2.2.2 What is the Process Context of the Proposed Experimentation?

You need to have an understanding of the process which your experimentation is attempting to improve. Part of this is knowing how that process sits within a context of other processes. This can often be conveniently displayed using a flow chart.

Thinking about the process context might prompt other questions. Perhaps neighbouring processes will be changed in the future. Perhaps there is also a need to improve neighbouring processes in some way, and the scope of your proposed experimentation could be made broader. Perhaps some effects of changing your process will be most obvious after subsequent processes have been completed. A Goal Hierarchy Plot might be useful for this purpose.

Example: Estimating parameters for modelling granular flows

A casual employee of CSIRO was asked to determine the frictional characteristics of iron ore particles of various sizes as they slid over a variety of surfaces. She was intending to make measurements using very fine particles as well as using particles with dimensions up to 30 mm.

The required task had been only loosely specified, largely because nobody was quite sure what to measure or how to go about measuring. Several people at a few different sites were concerned with the mixture of commercial reality, computer modelling and experimental measurements required to complete a project. The context could have been explained to the experimenter using the Goal Hierarchy Plot shown diagrammatically in Figure 1.

The top-level goal is to improve profitability of the metallurgical extraction part of some mining operations. Two of the ways of doing this are to reduce the running costs of mills and to improve the efficiency of the metallurgical processes used to extract the valuable components. Two ways of reducing running costs are to make better decisions as to when to reline mills and to reduce power consumption. A subgoal of improving metallurgical processes is to make the size distribution of the feed more consistent.

All three of the goals at the third level can be tackled by three approaches. The one of primary interest is to use a form of computational modelling called 'granular flow modelling' in which equations of motion are solved for individual particles.

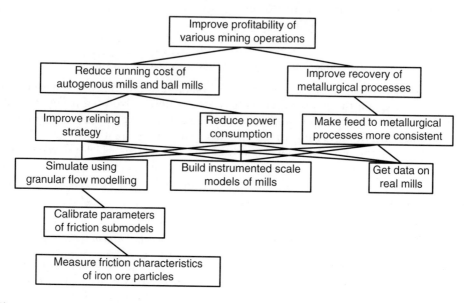

Figure 1. A Goal Hierarchy Plot to explain how granular flow models and experiments to calibrate such models fit into a plan to improve the profitability of some mining operations. The subgoals shown are useful for achieving the higher-level goals, but they are not the only things which are useful or necessarily the most important

In order for the computations to be feasible with the 1998 generation of computers, only particles larger than about 5 mm seemed likely to be modelled. Smaller particles might be considered by adjusting the characteristics of a continuous medium which is to be used as a proxy for the slurry of water and particles not otherwise modelled. Fine particles are not to be included in the model, so there is no need to determine their individual frictional characteristics.

For all this to have been explained would have helped the employee to understand the purpose of the experiments which she was performing, thereby avoiding doing experiments which were not worthwhile.

2.2.3 Have other People Tackled Similar Problems?

If similar problems have been solved, find out about the solutions. If the solutions are not public knowledge, find out what you can. If a solution has commercial value, someone should be taking advantage of it, perhaps even trying to sell it to you. If other people have been less than completely successful, find out as much as possible about what caused difficulties.

2.2.4 Have there been Prior Experiments?

There may have been prior experiments which have some relevance to achieving your objective. They may give insight into the reliability of the possible experimental

CLARIFY THE OBJECTIVE

protocols. You may be able to estimate the likely cost and precision of experiments which you might conduct. Perhaps prior experiments have made substantial progress towards solving your problem.

Perhaps your problem has been solved, but it hasn't stayed solved! Due to poor maintenance, changes in raw materials or compromises intended to reduce costs, something has become a problem. Perhaps reviewing the results of previous experiments might cast light on what should be changed.

2.2.5 Will there be Experiments after Yours?

I have seen several experiments in which people have (incorrectly) tried to solve a problem or to optimize a process while experimenting at the laboratory scale. They failed to realize that their objective was not to find a solution but rather to provide useful information to people who would subsequently experiment at a larger scale.

Usually, some aspects of the predictions of laboratory trials will remain reasonably accurate in full-scale application, but other aspects will not. Laboratory trials should often regard future larger-scale trials as their customer. They should find the interesting region and assess sensitivities so that unsatisfactory larger-scale trials are likely to produce useful results quickly.

2.2.6 What Outcome is Sought?

There may be assumptions which have been made about the process which are now being questioned. Perhaps there are new competitors threatening existing markets. Perhaps improvements are planned in order to develop new markets. Perhaps the goal is to fine-tune a process in order to make it more efficient or more robust.

It is always helpful to ask 'Who owns the problem?', 'Whose actions might be changed?' and 'What are the criteria for comparing the alternatives?'. A problem is only a problem insofar as somebody might be better off under some conditions, as yet not well known. If solving a problem would have no effect even in the very long term then the problem should be regarded as purely academic and ignored.

In situations where the consumers of a product or service would be better off if the product or service were improved, the producers of the product or the providers of the service might be regarded as owning problems associated with the quality of the product or service. They generally have a long-term interest in whatever affects their customers.

You must get some assurance from whoever owns the problem that they accept your formulation of the problem and might, under some conditions, be willing to implement a solution. Otherwise, don't bother.

In an industrial context, many problems are owned by the chief executive. Often some of the concern to implement solutions will be delegated.

2.2.7 What might be Done Differently?

When trying to clarify an objective, one very important key is to think very specifically about how the future might be different. Perhaps you are trying to influence a later stage of experimentation by someone else? Perhaps an operator might follow different procedures. Perhaps you wish to influence a capital expenditure decision.

The links between experimentation and implementation of improvements based on the information gained should be thought through as far as possible. An experiment which is perfectly designed for answering the wrong question may or may not be of any use.

Example: Experiment on raspberry canes

Someone wanted to do an experiment on raspberry canes to show that raspberries could be grown equally well from canes or roots. He had been told to consult me in my role as a biometrician about the the design of his experiment. We had a telephone conversation which proceeded something like this.

'*Why do you want to do this? Surely, such experiments have been done before? Why do you need to show this for local conditions?*'

'We are growing virus-free stock. The farmers are used to using canes for growing their next crop. We would like them to be willing to use root stock as well, so that we don't need to propagate as many plants under virus-free conditions.'

'*So the problem is not one of scientific research. Rather, the problem is to demonstrate results to the satisfaction of farmers. The standard theory about 'design of experiments' is not relevant to your problem. You may tell your boss that I told you:*

1. *You should put your experimental plots between the farm gate and the point of sale.*
2. *The most important measurement is the taking of large colour photographs every few weeks.*'

2.2.8 How General a Solution is Required?

Do you want a solution which will work in both summer and winter? Does it matter if your proposed solution creates slight hassles for other people either upstream (your suppliers) or downstream (your customers and consumers)? Does your proposed solution complicate procedures because special cases must be handled separately? If you solve the problem, will the solution be useful for products or processes other than the one which highlighted the problem?

What is the proposed range of application?

Clearly define the range of situations to which conclusions will be applied. Does it include all pieces of measuring equipment in some class, a range of machines, several products, several processes, all operators, all sources of reticulated water or other raw materials, all seasons or types of weather, all sites from some class, or all living things in some class?

CLARIFY THE OBJECTIVE

Will adjacent processes be modified?

It is useful to ask whether the context of nearby processes is likely to be changed in the near future. Changes to adjacent processes may have a different impact on some proposed solutions than on others. The economic impact of proposed changes may also be evaluated differently because of changes to adjacent processes

Will automatic process control be used?

One particular case of considering whether adjacent processes will be modified is to ask about automatic process control processes. If automatic process control is currently used for keeping a process under control then modifications to the process need to be evaluated very carefully. There are four possibilities.

- The existing automatic process control can still be used. The way of evaluating the proposed changes to the process must nevertheless take the automatic process control into account.
- The algorithm for automatic process control will need to be changed. This is complicated because a proposed change may fail to be an improvement if evaluated using the existing automatic process control algorithm.
- It may become possible to run the process without any automatic process control. Here there may be extra benefits, due to the cost savings of not needing automatic process control equipment.
- The process may become even less stable, so that an even more sophisticated automatic process control system is required. Perhaps tighter specifications on inputs will need to be introduced.

2.2.9 How will Outcomes be Evaluated?

Outcomes might be evaluated differently by different people. Perhaps likely future methods of evaluation can be anticipated to some extent.

- List the economic consequences of possible outcomes as quantitatively as possible. For instance, when considering possible ways of discouraging bank robberies, benefits of reducing the number of attempted robberies might include reduced emotional trauma for staff, a competitive advantage in attracting staff and a marketing advantage of greater perceived security compared to other banks.
- Remember that other parties' written specifications of their requirements may not correspond very well to their future requirements or even to their current requirements. If the output from your process is the input to someone else's process then their real requirement is that your output does not cause them any trouble.
  For instance, importers of iron ore require that a range of physical and chemical properties of the ore be within specified limits. However, if a new ore-body is mined which produces ore which satisfies the physical and chemical specifications but does not perform well in a blast furnace then a blast furnace operator can be expected to tighten up the formal specifications.

A cautionary remark: If a variable is adopted as a basis for payment or quality control then people will try to optimize that variable. Such behaviour will change relationships involving that variable because this is to their economic advantage. This is often a reason for doubting the generality of some proposed solutions to problems.

2.2.10 How Complex a Model is Acceptable?

It is often important to consider whether the complexity of a possible solution might be a barrier to its effective implementation. Complexity is acceptable for a new process which is expected to have high profit margins and to be operated by highly trained staff.

For processes which are far from new, and are often similar to those used by competitors, there is great economic pressure to automate these processes. Aspects of process complexity which can be handled by automatic process control are generally acceptable, but otherwise there is great pressure to simplify processes.

2.2.11 Are there any Environmental, Safety or Ethical Issues?

If there are any environmental, safety or ethical issues, these need to be raised as early as possible. Plans might need to be made to handle these issues.

Experiments on people generally require that the people undergoing experimental treatment be informed of any risks and that they give their consent. People should not be coerced to participate in experiments, but mild encouragement might be given provided that participants are not expected to suffer more than inconvenience or mild discomfort.

The ethics of experimentation on other species of animals is a more controversial topic, because much of this experimentation is not expected to have any benefit whatsoever to the species being experimented upon and because the animals cannot give 'informed consent'. The minimum acceptable level of ethical concern is to carefully consider how to cause the least pain and suffering while obtaining the desired amount of information. What have become known as the 'Three Rs' of Russell and Burch (1959) provide a logical basis for discussing animal experiments. These principles are replacement, reduction and refinement.

- **Replacement** means avoiding experiments on conscious, living, vertebrate animals by finding substitutes, such as experiments on tissue cultures, micro-organisms, insects, plants or anaesthetized mammals.
- **Reduction** means reducing the number of animals experimented upon, usually by refining experimental and data-analytical techniques.
- **Refinement** means reducing the distress of those animals which are used.

A practical interpretation of the question 'Are there any environmental, safety or ethical issues?' is to ask whether a planned programme of experimentation could withstand informed public scrutiny. For experiments on people, I believe that all aspects of the process of getting people to participate, but not personal details of participants, should actually be available for public scrutiny. Other types of experiments might be conducted without public scrutiny of the experimental protocols,

CLARIFY THE OBJECTIVE 21

perhaps because it is feared that a small number of people holding extreme views would interfere with the experiments, but people conducting such experiments should at least satisfy themselves that the experiments would not be considered unreasonable if a balanced, well-informed committee were ever constituted to review them.

Example: Preservation of grapes (continued)

Let us consider some of the above questions as they might be applied to the example of preservation of grapes which was introduced on page 12.

- There have presumably been prior experiments, because there is a suggested type of sulphur dioxide releasing pellet and a suggested size of holes in the plastic bags to hold the grapes. It would be worthwhile to find out exactly what experiments had been done and what the outcomes were. Perhaps there is no need to perform the planned experiment because the earlier trials produced a reliable conclusion. Perhaps the earlier trials were for a different type of grape being stored at a different temperature for a different period.
- There may need to be further experiments after the planned trial. Perhaps the performance of the two grape preservation methods varies with the maturity of the grapes or with the time between picking and arrival at the coolstore.
- The evaluation of the experimental results depends on the outcome which is sought. If we are considering whether to do more experiments and the use of pellets is more expensive than the standard procedure, then we are likely to proceed with more experiments only if pellets do a significantly better job of preserving grapes. If a supermarket chain and its associated coolstores were considering restructuring their handling of grapes then they would want to assess the cost and convenience of the alternatives as well as comparing the quality of the grapes from the final consumers' point of view.
- The preservation methods might be applied to other types of grapes also. Perhaps sultanas are being tried first simply because they constitute the largest segment of the market. Perhaps there are reasons to believe that the conclusions will not be applicable to some other types of grapes.
- Suppose that a method was better for grapes of one variety or harvested during one part of the harvesting season. This would mean that coolstores would need to have some sulphur dioxide in the atmosphere, in order to preserve those grapes for which this was the preferred method. The marginal cost of treating other grapes in this way would then be very low.
- The only environmental issue that I see as being important is the disposal of the plastic bags, but other people's opinions will be more useful than mine.
- The use of pellets may be slightly safer than having sulphur dioxide throughout a coolstore, but consulting a wide range of people may suggest additional safety issues.

2.2.12 *On the Manner of Asking Questions*

The manner of asking questions, such as those asked earlier in this section, can sometimes be as important as the questions themselves. I have three remarks to make about this issue.

- It is a good idea to ask open questions, inviting discursive answers, rather than closed questions. This will help open up discussion before aiming at specific goals.
- Do not be afraid of asking silly questions. If you feel concerned about appearing to be foolish or ignorant, think about how much more foolish or ignorant you will appear if you do not ask a question now but rather ask it at your next meeting!
- While asking questions, it is important to ensure that people understand each other's terminology. Sometimes technical terms or jargon can be a barrier to understanding, so be careful to clarify any possibly misleading terms. Encourage people to repeat ideas in their own terms in order to check that they understand them.

2.3 STATISTICAL MODELS AND TYPES OF EXPERIMENTAL OBJECTIVES

In this section, statistical models are described as including a trend for a response variable and a description of the likely departures from that trend. An experimental objective can be classified according to which parts of a statistical model you are trying to refine.

Many people working in engineering and the (hard) physical sciences see statistics as simply a minor tool which handles experimental errors. They tend to think that statistics is more useful in the (softer) biological sciences than in the hard sciences because the soft sciences do not have as good theories as do the hard sciences.

I believe that it is useful to note that in the hard sciences nearly all of the theories and models are concerned with explaining trends. Statistics is not essential for studying such theories and models, although it is often useful for fitting such models when the available data is affected by noise. Statistics is more useful when we are looking for a theory or a model to explain variability. The soft sciences have a larger proportion of theories and models that are concerned with variability. For instance, in genetics there are models for the amount of variation associated with genetic similarities between relatives and the amount of variation associated with environmental effects.

There are some situations in the hard sciences for which understanding the structure of the random variation is of primary importance. For instance, one aspect of the accepted theory for radioactive decay is that the number of counts usually follows a Poisson distribution. This is a very reliable and useful model, although it is a model for the variability rather than for the trend.

2.3.1 Statistical Models

In mathematical terms, suppose that we write a model as

$$y = f(X, \theta) + \varepsilon$$

where y denotes a response variable, X denotes the levels of some controllable or independent factors, θ denotes some parameters, and ε denotes random variation from the trend, $f(X, \theta)$. A person with a deterministic view of the world might focus on $f(X, \theta)$, and regard the random variation as always being random error which wouldn't exist if only the measurements could be done better. A person with a broader

CLARIFY THE OBJECTIVE

view might think about questions involving the distribution of ε, and whether this distribution could be changed.

Models which explain variability often distinguish two or more components of variation. The breaking of variation into components is useful because it provides insight about how total variation might be reduced.

If there is more than one response variable then we can think about them one at a time, possibly concentrating first on the one for which the most complicated model is thought likely to be required.

2.3.2 Types of Experimental Objective

Many types of experimental objective are possible. Here we consider which parts of a statistical model an experiment might help to estimate, test or otherwise refine. Four broad categories of experiments will be discussed.

First category: Main effects experiments

I will be using the term 'main effects experiments' for trials aiming to find simple descriptions of how factors affect a response. One simple type of description is a model in which the trend $f(X, \theta)$ includes main effects for the factors but does not include interactions and in which the departures from the trend, ε, are described by a single component of variation. Experiments which are likely to lead to slightly more complicated models, perhaps including a small number of interaction terms or describing the variation of ε using two or three components, will also be described as 'main effects experiments'.

Sometimes there are many possible factors and we suspect that only a small number of factors are important while most are irrelevant. We need to screen candidate factors in order to find which are important. In such cases, the main effects experiment is described as a 'factor screening experiment'.

In clarifying our objective, we need to consider whether it is more of a disaster to fail to find an important factor or to wrongly find that an unimportant factor appears to be important. Statistical theory talks about the relative importance of errors of types I and II. Errors of type I are considered to have been made when something is incorrectly regarded as significant. Errors of type II are considered to have been made when something is incorrectly regarded as insignificant when it is actually significant.

When experimental results are intended for publication in learned journals there is great emphasis placed on minimizing type I errors.

In medical trials, ethics demands that we do not use a treatment routinely unless it is effective; type I error is important. Recognising effective treatments (minimizing type II error) also has great economic importance.

For industrial trials, when experimental results are used for setting the parameters of processes, the careful specification of irrelevant factors is generally less important than a failure to recognize a significant factor. Type II error is the more important issue.

Second category: Response surface experiments

I will be using the term 'response surface experiments' for trials aiming to find more complicated descriptions of how factors affect a response. Such trials aim to fit models in which the trend $f(X,\theta)$ allows some interactions between factors. One commonly used mathematical form for the trend is a quadratic surface, such as $\theta_0 + \theta_1 X_1 + \theta_2 X_2 + \theta_{11} X_1^2 + \theta_{12} X_1 X_2 + \theta_{22} X_2^2$ which is linear in θ and has linear, quadratic and cross product terms in the components of X. As with main effects experiments, the departures from the trend, ε, are generally described by a single component of variation.

One typical example of a response surface experiment might be aiming to optimize a chemical process. It is common for most of the factors to be continuously adjustable.

Another typical example of a response surface experiment is checking the accuracy of a theoretical model. It is often useful to use response minus prediction as the variable to be empirically modelled in such experiments, which could be regarded as response surface experiments on the residuals.

Third category: Experiments studying sampling and testing errors

I will be using the term 'experiments studying sampling and testing errors' for trials aiming primarily at finding out about the typical size and structure of the random variations, ε. Interlaboratory trials are an instance of this type of experimentation which is likely to be familiar to most people.

One issue on which I would like to express a strong opinion is the importance of considering likely sampling errors. Sampling errors are generally more difficult to assess than testing errors, and are often given insufficient attention. I encourage experimenters to attempt to estimate the variance of sampling errors, as a step in improving the quality of their data.

Even when finding out about the size and structure of the errors is not the primary objective, it is often a secondary objective because an objective measure of error variance is required for statistical testing of the conclusions. In attempting to improve a manufacturing process, an experimenter is often not required to estimate the variance of items about the target to great precision because this is not important for deciding what actions to take. Also, routine monitoring of the process may have provided some information about the variance before experimentation begins. In contrast, if it is intended to publish the conclusions of experiments in scientific journals then an objective assessment of error variance is often required.

In animal and medical trials, experimentation must be able to assess effects of factors with enough information about components of variation to allow formal tests of statistical significance to be conducted.

In agricultural trials it is often necessary to estimate the intrinsic variability of the experimental plots for every experiment which is done. This is because soil is variable and it is much more variable in some places than in others. Plots which varied little in a year when there was little rainfall might be much more variable in a year when there were several downpours. Plots might be similar for one aspect of fertility but variable for another. Even if the same plots are used for various experiments at different times,

CLARIFY THE OBJECTIVE

it is seldom sensible to assume that the variation amongst them will be similar at the various times.

Fourth category: Experiments with multiple objectives

I will be using the term 'experiments with multiple objectives', for trials aiming both to fit a complex model for the trend $f(X,\theta)$ and to find out about the typical size and structure of the random variations, ε.

These four categories of experiments will also be used for distinguishing approaches to experimentation elsewhere in the book.

Some experimental studies involve several phases, with different objectives in the various phases. For instance, a first exploratory phase may consider many factors to see which are important. A second phase might systematically explore the effects of the most important factors. A third phase might then check that a theoretical model provides adequate predictive power.

A study of sampling and testing variation might take place at any time relative to the other phases.

2.4 TAGUCHI'S QUALITY ENGINEERING IDEAS

The ideas of Genichi Taguchi have been very fashionable. The ones that I am most interested in are sometimes described as 'quality engineering'. They can help us to think about the objectives of experimental programmes. The quality engineering ideas of Genichi Taguchi suggest that experiments be conducted as part of the design of products and processes in order to make these products and processes more robust. Taguchi's general philosophy of quality engineering which is explained in Kackar (1985), Taguchi (1986) and Taguchi (1987) considers product and process design to have three parts.

- **Systems design** is the process of applying scientific and engineering knowledge to produce a basic prototype design. This basic prototype will have many parameters which are yet to be specified. Initial guesses as to suitable values for these parameters are part of the systems design.
- **Parameter design** is the process of making a product or process less sensitive to identified noise factors. This is usually done by finding the settings for the parameters of the basic prototype design which optimize some performance statistic: usually a variance, a signal-to-noise ratio or the size of an operating window. The optimizing is not a trivial task and usually involves experimentation. You need to think carefully about noise factors and decide which to include in your experimentation and at what levels.
- **Tolerance design** is the process of deciding on tolerances for aspects of the manufacturing process and for components which will be used. Sometimes people assign tolerances considering only the capability of the manufacturing process and the variability of components from the least variable source, but this tends to keep costs high. Taguchi argues for rational assignment of tolerances based on estimates of costs and benefits, and for using experiments to find the sensitivity

of the performance to the variation in aspects of the product or process. He also argues that it is sensible to use a quadratic loss function – meaning that there is no loss if a characteristic is on target but a loss is considered to occur if the characteristic is off target, the loss being proportional to the square of the distance from target.

Taguchi's approach is particularly useful when there are many controllable factors which could be used to adjust a process average. Many manufactured products and manufacturing processes have this feature. Experimental effort can concentrate first on making the process more consistent, more robust or less sensitive, knowing that the process average can be readily corrected.

Another useful way of thinking about Taguchi's approach is that we want to be able to use tolerances which are as wide as possible. This includes both tolerances to environmental factors and to controllable process variables.

It is perhaps worth reminding readers that Taguchi should not be regarded as singularly important. See Box, Bisgaard and Fung (1988) and Goh (1993) for attempts to put Taguchi's ideas in perspective. Shainin and Shainin (1988, page 144) wrote

> The authors were surprised to learn, during a recent trip to Japan, using persistent and *intensive*, repeated questioning (through interpreters) to get beyond the party-line stories, that his [Taguchi] methods had only seldom been used in that country.

The following example illustrates how to apply the parts of Taguchi's ideas which I consider to be the most useful.

2.4.1 Example: Leaf Springs for Trucks

Consider a process for heat treatment of leaf springs for trucks which has been discussed by Pignatiello and Ramberg (1985). A leaf spring assembly is transferred from a high temperature furnace to a high pressure press which induces the camber (curvature) and is then immersed in an oil quench. The camber of a spring is assessed by measuring its free height.

It seems that tight control of quench oil temperature is difficult or expensive to achieve, but that it is required to achieve a tight tolerance on free height. There are two ways of reducing the effect of quench oil temperature variation on spring free height.

- Reduce the variation in quench oil temperature.
- Change the sensitivity of spring free height to quench oil temperature. This means asking that the slope of the relationship of the type illustrated in Figure 2 be reduced from the current slope. Figure 2 shows how the free height of the springs might depend on the quench oil temperature for two sets of conditions, A and B. For A, the free height is not sensitive to quench oil temperature. For B, the free height is sensitive to quench oil temperature. The average free height can be adjusted by changing the shape of a mould in the press or by other means.

If it is achievable, the second option is likely to be cheaper.

The experiment described by Pignatiello and Ramberg could be regarded as estimating the average slope of the graph under a number of conditions in order

CLARIFY THE OBJECTIVE

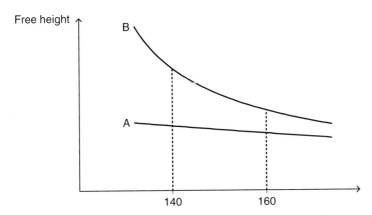

Figure 2. Two possible relationships of spring mean free height to quench oil temperature. The height is less sensitive to quench oil temperature for relationship A than for relationship B

Table 1. Differences in mean free height of springs between 140°F and 160°F quench oil temperature for some combinations of four controllable factors

Furnace temp. (°F)	Heating time (sec)	Transfer time (sec)	Hold down time (sec)	Free height difference (inches)
1840	23	10	2	0.34
1880	23	10	3	0.05
1840	25	10	3	0.54
1880	25	10	2	0.32
1840	23	12	3	0.00
1880	23	12	2	−0.01
1840	25	12	2	0.50
1880	25	12	3	0.34

to try to reduce it. The quench oil temperature is difficult to control precisely, so three experimental runs were done at 130–150°F (approximately 60°C) and three experimental runs were done at 150–170°F (approximately 70°C). The estimated differences in mean free height between quench oil temperatures of 140°F and 160°F are given in Table 1.

A glance at these experimental results suggests that it should be possible to reduce the sensitivity of spring free height to quench oil temperature by appropriately setting some of the parameters of the process. Both sets of runs with heating time in the furnace being 23 seconds and transfer time being 12 seconds seemed to be insensitive to oil quench temperature.

Pignatiello and Ramberg (1985) followed Taguchi's terminology and talked in terms of the signal-to-noise ratio. Some other people talk in terms of dispersion experiments which look at the effects of controllable factors on the noise variance. In my view, neither of these is as helpful as discussing the purpose of the experiment in terms of reducing sensitivity or improving robustness. The effect which is of interest to us can be described more simply by saying that we want a graph like Figure 2 to have a small slope than by talking about factors affecting the variance or talking about factors affecting the signal-to-noise ratio.

2.4.2 Differences between 'Research' and 'R&D'

Taguchi's view of the stages of product and process development illustrates some important differences between research and R&D.

- Research is more difficult to manage and plan, because you do not know how near you are to something of commercial value. In contrast, the value added by R&D should be assessable.
- Research can reasonably be done using high-quality materials and components. R&D should aim to use low-grade materials and components unless it is shown by tolerance design that higher-grade components or materials are economically justified.

3

Summarize Beliefs and Uncertainties

This chapter is about preparing a statement of what you know and what you don't know. Such a statement is a crucial milestone in any programme of experiments. In attempting to prepare such a statement, it is very common to change your view about the nature of the major problem, or at least to formulate some questions which might elucidate relevant issues.

3.1 DECIDE WHETHER YOU NEED AN EXPERIMENT

The previous chapter was about starting a programme of research. If you have followed the ideas in that chapter then you should have discussed some ideas with several people, raised lots of questions and made some progress towards agreeing on a clear objective.

This section is about deciding whether an experiment is needed, rather than, for instance, a survey or some analysis of routinely collected data. The remainder of the book is only relevant if experimentation is likely to be one of the ways of trying to reach your objective.

If you are intending to proceed further then you should be prepared to defend the belief that an experiment is justified. You should believe that it is possible, ethical and safe to deliberately vary the factors. You should believe that there is sufficient interest in the information which might be obtained to justify the likely expense and effort.

Example: Road maintenance strategies

Given that there are substantial differences in road maintenance practices between organizations in different regions of Australia, it seems likely that the potential benefits of obtaining good information about the effectiveness of various road maintenance strategies would be sufficient to justify research into road maintenance practices.

To decide on the best practices would require information about the costs and performances of various options under a wide variety of conditions. However, the currently available observational data on the effectiveness of various road maintenance practices does not provide the required information. There are a number of problems which are virtually impossible to avoid with this type of data.

- Pavement strength measurements are generally made infrequently, often only at the time of construction. It would be useful to relate pavement strength to changes in road roughness, but this would not be practical because it would require many more measurements of pavement strength.
- Traffic flow measurements are only approximate. Estimation of traffic loadings has improved considerably in recent years with the advent of classifier counters and weigh-in-motion systems, but much traffic loading data is still derived from axle counts and local estimates of heavy vehicle proportions and weights. It would be desirable to use a summary of the sum of fourth powers of axle loads or similar measures. Other problems with the estimated traffic loadings are that they actually vary with time of year and position along the road segments, but there is not adequate data about this variation.
- Climate and soil variables are not measured in sufficiently fine detail to provide much explanatory power.
- Ways of recording maintenance expenditure vary from region to region, and are therefore confounded with the differences in maintenance practices.

No doubt the quality of data will improve as years pass, but waiting for better data to arrive is not rational economic behaviour in my opinion.

I believe that experiments should be set up to directly investigate the questions of most interest and most economic importance. Some information about road maintenance practices in which loading is an important factor could be investigated using an accelerated load facility in which test sections of road were tested using controlled wheel loadings. A programme of experimentation could start by comparing maintenance practices similar to those currently used in various regions of Australia on various types of pavements.

Another useful type of experiment would be to compare different practices as directly as possible. The traffic, climate, soil type and road construction method should be essentially the same in adjacent experimental sections of road which could be maintained according to different trial regimes.

Other experiments investigating the merits of various maintenance practices would need more time to produce useful information. They would involve using different design, construction and maintenance practices on adjacent sections of road and monitoring the performance of these experimental sections of road over time.

3.1.1 What is an Experiment?

There are two features which distinguish experiments from other forms of data acquisition.

- An experiment involves deliberately varying factors in order to see what happens, rather than just waiting and watching to see what happens.
- Where there is more than one factor, an experiment should be nearly balanced in order to ensure that the effects of those factors can be separated. This allows the effects to be interpreted as causation.

So far as I know, the only meaningful interpretation of causation is in terms of a hypothetical experiment. For instance, when people say that cigarette smoking 'causes' lung cancer this can be interpreted as a prediction about a hypothetical experiment in which some participants chosen at random were required to smoke and the others were required not to smoke. Lung cancer would be more common amongst the smokers. (As in many other situations, a real experiment could not be conducted for ethical reasons.)

Taking a process view

It is often useful to take a process-oriented view of what is happening in order to generate ideas about how experiments might be carried out. Consider the following examples.

The chance that a bank will be robbed, or that a robbery will be attempted, depends on many characteristics of the bank – the suburb in which it is located, whether various countermeasures have been installed, the amount of pedestrian traffic, proximity to a police station, and many other characteristics. Experiments could be conducted to see how the frequency of attempted robberies varies with some of the parameters describing the bank.

The process of appointing new staff can be judged by how long the staff stay with the organization and measures of their promotion or effectiveness. Options within the appointment process could be investigated by systematic experimentation.

3.1.2 The Relative Merits of Observational Studies and Experiments

Experimentation has the good feature that the effects of possible causative factors can be separated. Factors included in an experiment are generally given sets of levels which are perfectly or very nearly orthogonal. Factors not included tend to be nearly orthogonal to factors which are included as factors because of the random assignment of treatments to experimental plots, animals, people or run sequences.

Observational studies generally have one important advantage over experiments. They are cheaper. Their primary disadvantage is that they are less conclusive. This inconclusiveness arises because the effects of possible causes cannot be reliably differentiated. For instance, observational studies have shown links between smoking and lung cancer and other causes of death for many years. Yet the idea that smoking is dangerous is not yet universally accepted. One reason for doubting that the observed association between smoking and lung cancer provides evidence of causation is that perhaps a genetic predisposition to lung cancer is correlated to another genetic characteristic which makes people more likely to enjoy smoking. Freedman (1992, pages 252–254) provides an overview and some links to the massive amount of research and controversy on this topic.

It has often been stated that observing a correlation does not imply that there is causation. The simplest way to think about this is that there may be a third variable which is affecting one or both of the variables which have been observed to be correlated.

People also need to consider the possibility of such third variables in order not to fall into the trap of assuming that lack of causation implies lack of correlation. For instance, suppose that analysis of some data suggests that there is a probability of 0.002 that the pilot of a light aircraft will fail to notice another aircraft, although the two aircraft will pass within a specified distance of each other if no evasive action is taken by either pilot. In estimating the probability of a near miss or a collision, it is tempting to argue as follows. Step 1: whether Pilot A sees aircraft B cannot have a causative effect on whether pilot B sees aircraft A. Step 2: hence the two events of visual detection are independent. Step 3: hence the probability that both pilots will both fail to see the other's aircraft is $0.002^2 = 0.000004$.

The problem with this logic is that there may be third variables which affect both pilots. Perhaps visual detection of other aircraft is particularly difficult at dusk, in particular weather conditions, over particular terrain or at particular angles of approach. The independence in step 2 of the logic above only applies for fixed values of the third variables, so step 3 is not valid. Arguments which calculate probabilities of failure for complex systems using probabilities of various types of partial failure should always be checked for this possible logical flaw.

The concept in probability theory which is most closely related to causation is not 'correlation' but 'partial correlation'. The partial correlation of A with B given C is the tendency of A and B to be similar which is not explained through the intermediary of C. For instance, the correlation of a characteristic of an animal with the same characteristic measured on a grandparent is often substantial; but the partial correlation given the same characteristic measured on the appropriate parent will be nearly zero. In theory, the genetic partial correlation of grandparents to grandchildren given parents is zero. This corresponds to the fact that there is no causative mechanism for a grandparent's genes to affect the grandchild other than through the genes of the parent.

Example: Health and cigarette consumption

Consider the artificial data set displayed in Figure 1. The horizontal axis shows age in years and the vertical axis shows a score indicating level of general health. The points plotted as triangles correspond to non-smokers. The points displayed on the graph as circles correspond to smokers. The areas of the circles are proportional to the average daily cigarette consumption reported by the women.

For this sample of women, the smokers are on average healthier than the non-smokers. If we had not realized that there was a third variable (age) which was correlated with both smoking and health then we might have got the wrong impression.

Once the third variable has been identified there are several ways of allowing for it in the data analysis. For instance, we could fit a model which assumes that health score is affected by both cigarette consumption and by age, or we could look at the relationship between age-corrected health scores and cigarette consumption.

3.2 CONSIDER RELEVANT SCIENCE AND TECHNOLOGY

Relevant scientific and technological information should be sought as part of your review of beliefs and uncertainties. Such information might be found in open technical literature, possibly with the assistance of a librarian. More specific technical information might be found using your organization's corporate memory, in whatever form that takes, and using technical information from sources such as manuals

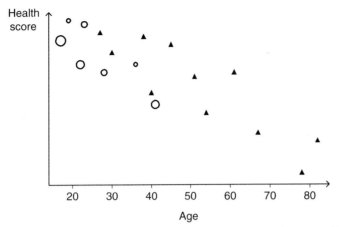

Figure 1. Plot of health score against age for a number of women. Triangles correspond to non-smokers. Circles correspond to smokers, with area of circles proportional to cigarette consumption

describing testing equipment, national or international standards, colleagues and collaborators.

In particular, check whether there are models available. Also, check whether there is work currently being done elsewhere in the world which is related to your proposed experiments. If a model is available, then note that response surface experiments can be made more efficient by using the discrepancy between a model and observed results as the response variable, rather than simply using the observed results. It is more efficient because the model explains much of the variation between results.

For factor screening, theoretical knowledge is generally of less use, but it can be used as a guide when interpreting results. An estimated effect which seems to act in the expected direction is more likely to be a real effect than if it seems to act in the opposite direction.

Example: A basic oxygen steelmaking furnace

When iron ore is reduced to iron in a blast furnace, the resulting pig iron contains a few per cent of carbon. The carbon needs to be almost entirely removed in order to convert the iron into steel.

One current mode of operating a vessel for converting pig iron into steel involves some pre-treatment of the iron and adjustment of the materials added to the iron according to its measured chemical composition and temperature. This is sometimes described as 'feed-forward'. The state of the refractory lining of the vessel and the amount of scrap iron might also need to be considered as part of the feed-forward model. There is also feedback based on measurements made during the process of blowing oxygen through the molten metal. It would be useful to refine the parameters which describe both feed-forward and feedback.

It would be foolish to attempt to analyse historical data on this process or to conduct experiments without considering the models of the process which are currently used for deciding how the vessel should be charged and for predicting the outcome of further blowing.

3.2.1 Knowledge about Variability

Poisson variation of radiation counts

One situation where most people are comfortable that they have reliable and useful knowledge about variability is in the variation of radiation counts. Such counts follow Poisson distributions. This means that if the average count is λ then the probability that the actual count will be x is

$$\frac{\lambda^x e^{-\lambda}}{x!}$$

and the variance of the counts will be λ. The fact that the variance is equal to the average allows the precision of counts to be estimated very simply.

This knowledge about variability can also be applied to X-ray fluorescence machines and many other radiation-based measuring devices.

3.2.2 Knowledge about Effects of Factors on the Mean

In many circumstances, people have quite a substantial amount of knowledge and opinions about the effects of factors on the mean. This information may be difficult to use because it is difficult to specify how reliable it is.

Sometimes people are reasonably confident about the direction of an effect. For instance, using more finely ground raw material would be expected to increase the rate of chemical reactions, because of the increase in surface area. This in-principle increase in reaction rate may turn out not to affect the practical reaction rate because of other constraints, but you might feel quite confident that there should not be a decrease in reaction rate.

Sometimes people have confidence about the signs of quadratic terms. For instance, a crop yield may be likely to be increased by the use of nitrogenous fertilizer. There will be an application rate for the fertilizer such that any increase beyond that rate has little effect or even decreases the yield. These remarks imply that it is expected that the second derivative of the trend in yield as a function of application rate will be negative.

3.3 SAMPLING OF PARTICULATE MATERIALS

This section presents some theory which may be useful for predicting the amount of variation associated with sampling. It is more mathematical than the rest of this book. Many readers may prefer to skip the section on first reading. I believe that it is relevant to much experimental planning and should be more widely known. Many organizations, including research institutes, have excellent measurement and testing facilities but poor sampling facilities and poor knowledge of sampling theory and procedures.

First, a few terms need to be introduced.

SUMMARIZE BELIEFS AND UNCERTAINTIES

Lot: a lot is a defined quantity of material from which a sample or samples need to be taken, possibly for the purpose of determining some aspect of quality or grade.

Sample: a sample from a lot is a quantity of material smaller than the entire lot which has resulted from a series of operations intended to make the small quantity of material representative of the entire lot.

Representative sample: a sample is representative if measurements made on the sample are likely to be similar to the same measurements made on the entire lot. This usually requires that all parts of the lot being sampled must have the same probability of becoming part of the final sample on which measurements are made.

Often, a carefully prepared sample will only be representative for some purposes, not for all purposes. For instance, a sample which has been crushed and dried might be representative for dry chemical analysis but not representative for moisture determination or considering particle size.

One component of sampling error is unavoidable. This has been referred to by Gy (1982) and Pitard (1989) as 'fundamental error'. It arises from the fact that the individual particles vary. Most variation between samples is due to the variation between the larger particles in samples which might be selected.

A stand-alone derivation of the sampling error variance is given below. The derivations in books that might be cited are difficult to follow. This derivation provides some insight as to how sampling variation arises.

3.3.1 Theoretical Derivation of Sampling Error Variance

Consider a lot of material consisting of n particles that we know everything about. The ith particle has mass m_i and grade g_i. The total mass of the lot is

$$M = \sum_{i=1}^{n} m_i,$$

and the average grade is

$$G = \sum_{i=1}^{n} m_i g_i / M.$$

Consider taking a sample which is approximately a proportion p of the entire lot. The influence of the inclusion or exclusion of particle i on the sample can be readily calculated by considering two possible constitutions of the sample, assuming that the remainder of the sample does not vary and is precisely a proportion p of the lot apart from particle i, therefore having mass $p(M - m_i)$ and average grade

$$\frac{GM - m_i g_i}{M - m_i} = G + \frac{m_i(G - g_i)}{M - m_i} \approx G + \frac{m_i}{M}(G - g_i).$$

The fractional error in the approximation is of order m_i/M which is negligible provided that the sample is large compared to single particles.

With probability $1 - p$, the sample will exclude particle i, and the sample will have the average grade given above. With probability p, the sample will include particle i.

In this case the sample will have mass $p(M - m_i) + m_i$ and grade

$$\frac{p(GM - m_i g_i) + m_i g_i}{p(M - m_i) + m_i} \approx G + \frac{(1-p)m_i}{pM}(g_i - G).$$

The mean grade considering these two possibilities is approximately G and the average squared departure of the grade from G is approximately

$$(1-p) \times \left[\frac{m_i}{M}(G - g_i)\right]^2 + p \times \left[\frac{(1-p)m_i}{pM}(g_i - G)\right]^2 = \frac{(1-p)}{p}\left(\frac{m_i}{M}\right)^2 (g_i - G)^2.$$

The total sampling error variance, V_S, is the total of such contributions to the sampling variance for all individual particles. It is therefore

$$V_S = \frac{(1-p)}{p} \sum_{i=1}^{n} \left[\left(\frac{m_i}{M}\right)^2 (g_i - G)^2\right].$$

The average mass of the sample is pM, which we will refer to as the mass of the sample and denote by M_S. The sampling variance can therefore be written as

$$V_S = \frac{(1-p)}{M_S} \sum_{i=1}^{n} \left[\frac{m_i^2}{M}(g_i - G)^2\right].$$

3.3.2 Interpretation and Use of the Formula for Sampling Error Variance

The quantity $\sum_{i=1}^{n}\left[(m_i^2/M)(g_i - G)^2\right]$ can be interpreted as the heterogeneity of the lot of material being sampled. It is much more sensitive to the grade of large particles than to the grade of small particles. It does not depend on the size of the lot. For instance, if a lot were doubled in size then there would be twice as many terms in the summation and M would be doubled in size so the quantity would be unchanged.

Note that the sampling variance is inversely proportional to the mass of the sample. One good way to estimate the sampling error variance in practice is to take small samples of blended material and to test them as accurately as possible. Because these samples are very small, sampling error variance should be much larger than testing error variance. Hence the variance of the test results provides an estimate of the sampling error variance for small samples which is almost unaffected by testing errors. The known (inverse) relationship between sampling variance and sample mass can then be used to calculate an estimate of the fundamental sampling error for the sizes of sample that will actually be used.

It is a common mistake to think that sampling error variance is strongly dependent on the proportion p; it is not. The sampling proportion p is generally small, so $1 - p \approx 1$ and V_S depends principally on the size of the sample and on the heterogeneity of the material to be sampled. For instance, the sampling error incurred when taking a 10 kg sample from a 300 kg lot is likely to be of similar magnitude to the sampling error incurred when taking a 10 kg sample from a 200 000 tonne lot. Although the sampling errors are likely to be of similar magnitude, the many other things that can go wrong when taking samples are more likely to go wrong when a large quantity of material must be handled. These other things are sometimes called 'non-sampling errors'. See the discussion of sampling bias on page 127.

SUMMARIZE BELIEFS AND UNCERTAINTIES

When the particles can be regarded as consisting of k classes then, provided that the sampling proportion p is small,

$$V_S \approx \frac{1}{M_S} \sum_{i=1}^{k} p_i m_i (g_i - G)^2$$

where p_i is the proportion by mass of particles which are in the ith class, m_i now denotes the mass of a typical particle in the ith class and g_i now denotes the grade of a typical particle in the ith class. From this formula we can see that coarse particles generally have the largest effect on V_S.

The formula can be rewritten to say how large a sample is required in order to achieve a desired sampling error variance

$$M_S \approx \frac{1}{V_S} \sum_{i=1}^{k} p_i m_i (g_i - G)^2.$$

Example: Blending of powders

A mixture of powders consists of 88% by weight of component A, 10% of component B and 2% of component C. Suppose that typical particle sizes are 100 μm for A, 60 μm for B and 40 μm for C. Taking the density of all three components to be 2 g/cm^3, we will estimate the standard deviation of the percentages of B and C in 250 mg samples of the mixture.

Taking the particles to be cubic in shape, typical masses of individual particles are 2 μg for A, 0.432 μg for B and 0.128 μg for C. For the percentage of B in a 250 mg sample,

$$V_S \approx \frac{1}{M_S} \sum_{i=1}^{k} p_i m_i (g_i - G)^2$$

$$= \frac{1}{250} \left[0.88 \times 2(10)^2 + 0.1 \times 0.432(90)^2 + 0.02 \times 0.128(10)^2 \right] = 2.1$$

Taking the square root of this variance, the standard deviation of the percentage of B is approximately 1.45%.

For the percentage of C in a 250 mg sample,

$$V_S \approx \frac{1}{250} \left[0.88 \times 2(2)^2 + 0.1 \times 0.432(2)^2 + 0.02 \times 0.128(98)^2 \right] = 0.127.$$

Taking the square root of this variance, the standard deviation of the percentage of B is approximately 0.36%.

Example: Sampling of gold ore

Suppose that the largest gold particles in a gold ore are of a size equivalent to spheres of diameter 4 mm, that the average grade is about 3 g/t or 3 parts per million (ppm), that the density of the rock is 2.7 tonnes per cubic metre, and that the required precision is a standard deviation of 0.6 g/t. How large a sample is needed?

If we regard the ore as consisting of 4 mm diameter spheres of two types, then we can apply the formula above. We will use gold grade in ppm as the quality characteristic. The gold particles ($i = 1$), will have particle mass of about 0.64 g (this being the mass of a sphere of radius 4 mm and density 19 tonnes per cubic metre) and grade of 1 000 000 ppm. The other particles will have particle mass of about 0.09 g and grade of zero ppm.

Applying the formula above, we calculate the sample size required to be

$$M_S = \frac{1}{V_S} \sum_{i=1}^{k} p_i m_i (g_i - G)^2$$

$$= \frac{1}{0.36} \left[0.000003 \times 0.64(999997)^2 + 0.999997 \times 0.09(3)^2 \right] \text{ g}$$

$$= 5.3 \text{ t}.$$

This is only an approximation, since the particles of gold are not really of uniform size. The material other than gold does not actually consists of spheres of 4 mm diameter, but the second term in the formula makes little contribution to the total so this assumption is less critical.

Example: Precision of sizing

This example is similar to example 18.4.3 of Pitard (1989). For coarse ore which can be regarded as 95% by mass of approximately spherical particles 140 mm in diameter and 5% by mass of approximately spherical particles 250 mm in diameter, assuming a density of 2.7 tonnes per cubic metre, how large a sample of material is required so that the proportion retained on a 150 mm sieve will have a sampling standard deviation of 0.5%?

There are two classes of particles ($k = 2$). For the 140 mm diameter particles ($i = 1$), typical particle mass is $m_1 = 3.88$ kg and the typical grade is $g_1 = 100$ for the grade characteristic 'percentage retained on a 150 mm sieve'. For the 250 mm diameter particles ($i = 2$), typical particle mass is $m_2 = 22.1$ kg and the typical grade is $g_2 = 0$ for the grade characteristic 'percentage retained on a 150 mm sieve'. Now the required standard deviation is 0.5, so the required V_S is $0.5^2 = 0.25$, and

$$M_S = \frac{1}{V_S} \sum_{i=1}^{k} p_i m_i (g_i - G)^2$$

$$= \frac{1}{0.25} \left[0.95 \times 3.88(5)^2 + 0.05 \times 22.1(95)^2 \right] \text{ kg} = 40.2 \text{ t}.$$

Such a calculation might be important for an experiment in which crusher settings were adjusted.

3.3.3 Test Precision Constrained by Sampling Errors

There are many test methods for which a large component of what might be described as testing precision is really a sampling problem.

Here are two examples.

- Near infra-red reflectance methods are only affected by the small mass of material near to the surface over a small region.
- A physical test for coarse iron ores heats lumps of ore in a reducing atmosphere and measures the mass of fine particles which break off the lumps. This test has been found to have poor reproducibility, largely because it takes only a small number of lumps of ore for testing.

3.4 CONSIDER INFORMATION FROM EXISTING DATA

Use exising data to try to relate response variables to controllable factors, but be cautious about the quality of the data and about the continuing applicability of such relationships. Any previous experiments should be reviewed.

The approximate accuracy and precision of sampling and measurement systems should be estimated using available data at this stage of the planning of an experiment.

No experiment is ever done without being in a context from which information can be extracted. At the very minimum, it is generally possible to get some idea about the precision of testing methods which might be used. Commonly, we have some idea about the likely direction of the effects of factors and some idea about what size of difference would be important. For some experiments, we can make use of information obtained from monitoring similar processes over a long period.

Example: Density of asparagus plants

Bussell et al. (1997) reviewed 15 yield-density studies for transplanted asparagus, and came to the conclusion that 'new research will benefit from the framework of a coherent understanding of past research, supported as far as possible by quantitative analysis'. They found that researchers had consistently chosen ranges for planting densities such that the optimum density was found to be at or above the largest density used in experiments. They also found that some studies used soil types different from those used in commercial practice.

Even if a new cultivar were to be tested, it is clear that such experimentation would benefit greatly from study of past experiments. This would help with deciding the density range and the soil types to be tested. It would also help with decisions about experimental protocol, such as the need to report climatic data, soil chemistry, fertilizer use and manure use; and with decisions about what models to fit to the data.

3.4.1 Looking at Existing Data as a Source of New Questions

The first phase of looking at existing data is often exploratory. In attempting to make sense of data I often find myself asking questions like the following.

- How, why and where were the measurements made?
- What are the units of measurement?
- How frequently were measurements made?
- Are the measurements continuous or discrete?

- How were measurement devices calibrated?
- Does the data correspond to what is measured or have some calculations been done?
- Should the numbers provided to me be regarded as independent pieces of data, or are some of them calculated using models?
- Is some information missing? Might estimates have been substituted for missing data?
- Were standard operating practices followed? ... Really! ... What happens on night shift? What would happen if ... ?
- Can the patterns and exceptions that I can see from plots of this data be explained by my current knowledge?

The steps that I follow during such exploratory data analysis might be listed in the following order of complexity.

1. For each variable in turn, look at the name, meaning and a few typical values. I check that I understand the processes which have been followed in order to measure or calculate this variable.
2. Look at all of the values for one variable at a time. What is the distribution? What are the limits? Do you understand how the variable changes over time?
3. Look at scatter plots for pairs of variables that might be related in some way. Can I understand how the relations apparent in the data might have arisen?

I use these steps primarily to help me to think of more questions to ask, with extracting information from the data being a goal of less immediate importance.

3.4.2 Getting some Information about Components of Noise

Routinely-collected data is almost always of some value for estimating the amount of variability to be expected.

- It lets us estimate the variability of process inputs.
- It lets us estimate the variability of process outputs.

Even small amounts of historical data are useful for these purposes, but we should always be aware of the possibility that long-term variation may exceed short-term variation.

- Historical data may enable us to see whether successive items are independent, to estimate the pattern of correlation, or to estimate several components of variance.
- Process capability studies tell us about the variation which is likely to occur in the future. It is useful to know about process capability studies for the inputs to a process on which you are experimenting.
- Measurement studies tell us about sampling and testing precision. Such information can be very useful. If sampling and testing precision do not appear to be adequate for your purposes then it may be necessary to improve them or to abandon your original goal.

Separately estimating several components of variance usually requires large amounts of data. Furthermore, it is often true that historical data contains no useful information about some of the relevant components of variance. For instance, long-term variation in process characteristics cannot be differentiated from long-term drifts in sampling and measurement processes.

The Ina tile experiment (see page 181) is a good example of the use of historical data to provide an error variance.

Statistical Process Control

In some segments of manufacturing industry, particularly suppliers to the automotive manufacturers, it is common to use some procedures which are known as Statistical Process Control (SPC). It would be more accurate to describe these procedures as Graphical Process Monitoring, but the term 'SPC' is in common usage.

The purpose of SPC is to detect changes from the normal or typical pattern of variation of a process. At the first stage of SPC, some assessment must be made of the amount of variation in the process which is normal or typical. Such an assessment is often very useful as a source of information about the variability likely to occur when experiments are conducted. Any indication that the variability has changed can also be very useful – it may be relevant to our choices of factors to investigate.

3.4.3   Getting some Information about Effects of Factors

In principle, historical data may be useful for estimating the effects of factors of interest, provided that these factors have been varied in the past, sampling and testing have been conducted consistently, and record-keeping has been good.

In practice, historical data is seldom of as much use as might be hoped for estimating the effects of factors of interest, because other factors have also been varied over the period of the records and short-term process adjustments have been made. These tend to disguise the effects of interest. However, some useful information can usually be obtained.

Absorption

Sometimes historical data will be affected by seasonal variation, batch variation or variation between individuals, as well as being affected by some factors that you might be interested in investigating. In such cases statistical analysis of the historical data might involve fitting a linear regression model with a parameter for each different season, batch or individual. One technique for doing this efficiently is often called 'absorption' (at least by people who work in parts of agriculture and genetics). It amounts to doing a Gaussian elimination of equations which are not of interest as soon as all information relevant to the those equations has been processed.

3.4.4   How much Information is there in Historical Data?

Many people imagine that because they have lots of data there must be a lot of useful information in that data. This is not necessarily true. The amount of useful information is just as dependent on the purpose for which the information is sought as it is on the database.

For instance, suppose that I had a database which described all telephone calls made between Europe and the USA over the last 15 years, and that I wished to predict the number of transatlantic telephone calls which will be made next weekend.

The database is very large. However, despite the amount of data available we cannot predict very accurately because the data and the quantity to be predicted are affected by many sources of variation. The causes of fluctuations in the rates at which people make telephone calls include time of day, day of the week, daylight saving time, the weather, major sporting and cultural events, holidays, religious events, special discounts on telephone calls and advertising.

3.4.5 Data Selection as a Source of Distortion

Observational data should always be treated with caution. However, selected data should be regarded as hazardous.

Consider the following possible distribution of children in a set of 100 families. (For simplicity, details of these families are considered to be available to us without any of the usual real problems that would be encountered when acquiring such data. None of the children or parents have died. The pairs of parents are all still together. There are no half sibs or children of uncertain parentage. People always respond to questionnaires, know the truth and report the truth. And none of the families will subsequently have any more children.)

Number of boys	Number of girls				
	0	1	2	3	4
0	12	12	8	2	1
1	12	16	6	4	–
2	8	6	6	–	–
3	2	4	–	–	–
4	1	–	–	–	–

In these 100 families, there are

$$(12 + 16 + 6 + 4) + 2(8 + 6 + 6) + 3(2 + 4) + 4(1)$$

$= 38 + 40 + 18 + 4 = 100$ girls. They have a total of $38(0) + 40(1) + 18(2) + 4(3) = 88$ sisters and the same number of brothers.

Suppose we selected some families by speaking to a set of girls and asking them about the families of which they are children. On average, such families would have 1.88 girls and 0.88 boys.

Similarly, if we selected some families by speaking to a set of boys and asking them about the families of which they are children, then such families have an average of 0.88 girls and 1.88 boys.

If we asked a collection of people about three quarters of whom were male about the families of which they are children, then such families have an average of 1.13 girls and 1.63 boys. This is contrary to what many people imagine.

Note the bias between asking adults how many children they have (the average is exactly 2) and asking children how many children are in their family (the average is 2.88). This might partly explain the feeling that families are getting much smaller.

When we are children we estimate average family size based on the families of our friends, who are children; whereas when we are adults we estimate average family size based on the families of our adult friends, who may or may not be parents.

Some common examples of selected data are the following.

- Mining the regions which appear to be high grade shows that unbiased estimates were too high on average.
- During wars, it is normal practice for news media to be selective as to which stories they report, perhaps because of only being allowed access to restricted parts of the combat region. The impressions given to the public by such selective reporting are usually biased.
- Tidying up data (deleting or correcting the data which doesn't fit your preconceived opinions) can easily make that data more supportive of your preconceived opinions.
- In meta-analysis, where many studies relevant to a particular issue are reviewed, someone with a strong personal opinion about an issue such as fluoridation can be particularly painstaking in reviewing studies which do not support their opinion. They are likely to find that a summary of the studies with acceptable methodology then agrees with their personal opinion.
- You always seem to catch up to speeding vehicles at the next red traffic light – at least the ones that you see again after they have overtaken you.
- Comparing the lifespans of unmarried men and married men always shows that married men live longer. Some of this effect may be due to married men being well looked after, both physically and emotionally, by their wives. However, some of the effect may be due to male children who die as babies generally being unmarried. Also, women may tend to choose as marital partners men who are more likely to have long lives.
Widowers live even longer than married men. That is to say, if we consider men who are married at the time of death of either themselves or their wives, then the men who outlive their wives on average live to a greater age than the men who die before their wives.
- Parents of a baby which sleeps regularly through the night at seven weeks of age tell all their friends. Parents of a baby which hardly ever sleeps through the night until two years of age keep this information to themselves.
- A telephone survey of voting intentions can only consider people who have telephones and who were at home when efforts were made to contact them.

Example: Survey bias

As discussed by Tanur *et al.* (1972, pages 146–7) and Wallis and Roberts (1956), one of the most famous examples of bias in survey data concerns the 1936 presidential election in the USA. A magazine called the *Literary Digest* based its predictions on a very large number of questionnaires, namely 2 376 523 returned out of about 10 000 000 mailed, but it had drawn its sample from lists of owners of automobiles and telephones. The predictions were wrong by 19%. It failed to predict that Roosevelt would win and looked very foolish when he won with a landslide.

With hindsight, it is easy to see that voting opinions were highly correlated with economic status at that time and that the sample take by the *Literary Digest* was not representative of the population of voters. The same method of sampling had correctly predicted the winners of some previous presidential elections for which voting behaviour was not related to income.

Example: Aircraft survivability

Contributing to the United States's military efforts during the Second World War, Abraham Wald worked on the problem of estimating the vulnerability of aircraft to enemy weapons. This work is reported in Mangel and Samaniego (1984).

A major complication of this work is that data is not available on the aircraft which did not return. For instance, the proportion of single-engined aircraft which had been hit in the engine can be expected to be much larger amongst the aircraft which did not return than amongst the aircraft which did. In order to allow for the selection bias in the data, a substantial amount of model-fitting work was required.

3.4.6 Incentives as a Source of Distortion in Data

When people know what is being measured and they have some incentive to make some measurements look good, it is common for those people to modify their behaviour in order to do so. See Dickinson and Robinson (1994) for a discussion of this issue.

The management of a company which provides installation and maintenance services for its products measured delays between requested service times and actual service times. A reaction from some servicemen was to manipulate the data. Requested service times could be adjusted by recording a customer's second preference for service time if the customer's first preference was unlikely to be achieved. Actual service times could be adjusted by starting jobs at the required times, even if this meant that jobs were not completed during that visit to a customer's premises.

Performance appraisal systems always provide data which is affected by incentives. If you tell people the criteria according to which they will be remunerated, promoted and respected then you can expect that they will improve the numerical measures of those criteria. If the performance appraisal system is itself evaluated by looking at changes in the average values of some criteria then it is to be expected that these averages will have improved, but this should not be interpreted as evidence that the appraisal system is working.

Critical evaluation of data affected by incentives is difficult. The only easy-to-remember name that I have ever heard for it is 'applying the greengrocer test'. The basis for this name is the idea that a greengrocer will always ensure that the fruit and vegetables which you see (and might think are indicative of average quality) are of better than average quality.

3.4.7 Attempts to Control a Process as a Source of Distortion in Data

Routinely collected data may be affected by efforts which are made to control processes. Sometimes this effort is made by operators. Sometimes it is made by automatic process control devices. Such effort can be extremely effective at hiding the real relationships between variables.

SUMMARIZE BELIEFS AND UNCERTAINTIES

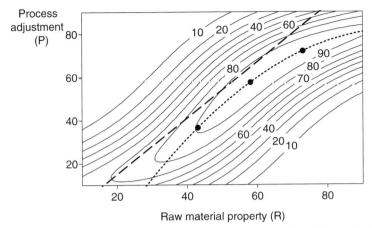

Figure 2. Contour diagram of relationship of average Y, to R and P. The continuous lines are contours at spacings of 10 units in Y. The dashed line shows one possible way of setting a process control regime in which P is adjusted according to the value of R. The dotted line shows a better way of setting the process control regime

Artificial example of process control

Consider a batch process in which an output quality characteristic, denoted by Y, is affected by a raw material property, R, and by a process adjustment setting, P. Suppose that the real relationship of the average Y to R and P is as shown by the contours on Figure 2. The distribution of actual values of Y about that average is assumed to be normally distributed with standard deviation 3.

We can see from Figure 2, by considering a transect parallel to the P axis, that for a fixed value of R there is a best value of P and that Y decreases as P is further from the best value. Similarly, by considering a transect parallel to the R axis, we can see that for a fixed value of P there is a best value of R and that Y decreases as R is further from the best value. Figure 2 also shows two lines describing possible process control regimes.

Figure 3 shows some data generated assuming that the property of the raw material, R, is normally distributed with mean 58 and standard deviation 10, and that the process control variable P was always set according to the value of R and the relationship shown by a dashed line in Figure 2. The dashed line shows how the average value of Y would be expected to vary with R, given the true relationship as shown on Figure 2.

This possible historical data does not tell us anything about the relationship between Y and R for fixed P, which is what most people think about when they talk about the relationship between a pair of variables when other variables are known to also be influential. Rather, the historical data tells us about the relationship between Y and R when P is adjusted with R according to a specified regime.

Similarly, Figure 4 tells us about the relationship between Y and P when there is the same regime for adjusting P according to R. This is not the relationship between Y and P for fixed R.

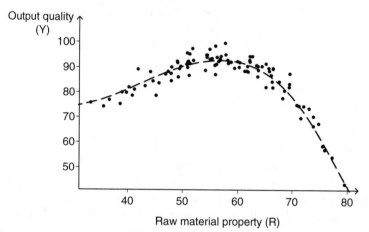

Figure 3. Possible historical data showing relationship of Y to the raw material property, R. The dashed line shows the average relationship

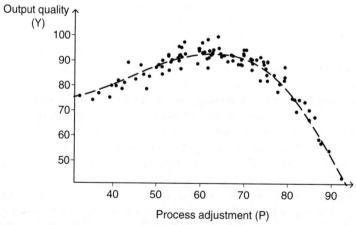

Figure 4. Possible historical data showing relationship of Y to the process adjustment setting, P. The dashed line shows the average relationship

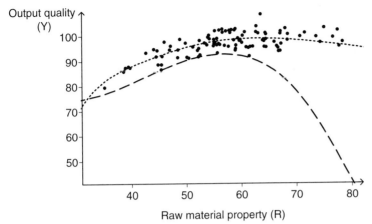

Figure 5. Relationship of Y to R when an alternative process control scheme is used. The dotted line shows the average relationship for the alternative process control scheme. The dashed line shows the average relationship for the original process control scheme

Both Figures 3 and 4 only tell us about the process for the subset of values of P and R which are shown on Figure 2 by the dashed line.

If we want to improve the process control scheme we need to get some information about how the process might perform if different process control schemes were adopted. The historical data shown in Figures 3 and 4 is of absolutely no value for considering changes to the process control scheme.

One example of some data which would be of substantial value is to know how Y varies with P for some fixed values of R. This requires running the process differently from the usual process control scheme. For instance, at $R = 43$ the usual process control scheme would set $P = 44.9$. A small experiment in which different values for P were tried would be likely to quickly demonstrate that a smaller value for P would lead to a higher value of Y.

Suppose that three similar experiments showed that good values for P were 36.5, 57.5 and 72, for R being 43, 58 and 73, respectively. A quadratic curve through these three points is shown on Figure 2 as a dotted line. Using this line to define an alternative process control scheme, Figure 5 shows the results of a process simulation with other features as in Figure 3.

The performance of the process has been dramatically improved. The average value of the response variable Y is much higher and the process seems to be more robust to the quality of the incoming raw material, R. Another desirable feature of the new control regime is that the process would be less affected by errors in the measurement of R.

From Figure 2, it can be seen that the process could be further improved by adjusting the process control scheme so that even lower values of P were used for low values of R.

In practice, things aren't usually this simple. There may be feedback and feedforward process control, several uncontrollable but measurable input variables, several process variables, substantial measurement error variances, correlations – and the process might not be stable. None of these problems has a great tendency to make historical data any more useful, and that's starting from a situation in which historical data is useless!

Be cautious. Remember that some effects have been taken away by adjustment. Also, you need information about the precise process control mechanisms which have been used.

3.5 LIST RESPONSE VARIABLES AND FACTORS

List the response variables which are of interest, whether or not you plan to measure them. Response variables may usefully include measures of variability, sensitivity or robustness. Consider the precision, timeliness, cost and relevance of these variables. Indicate typical values and any customer requirements.

Do different response variables have markedly different importance for different customers? If not, might it be possible to combine response variables into an overall utility measure?

List the factors which you think might influence the response variables. This might be based on a brainstorming session involving a wider group of people than the project team. Indicate which factors are easy to vary and which are difficult or expensive to vary. Be sure to include factors which might be necessary to ensure that conclusions made can be applied with sufficient generality. Indicate normal settings and likely alternatives.

Within a programme of experimentation, it can happen that a quantity is regarded as a factor for some trials but is regarded as a response variable for some other trials. This is most likely to happen when seeking to improve the overall efficiency of a set of processes.

The output from one process is an input for a subsequent process. An experiment on the first process could investigate how its inputs affect its output. An experiment on the subsequent process could investigate how its inputs (the previous output) affect its output.

In thinking about what quantities might be varied, controlled, measured or monitored, we will not yet be concerned about which of these quantities should be considered to be response variables and which should be considered to be factors. That will be discussed in Chapters 4 and 5.

3.5.1 Looking for Response Variables

My thinking in this area has been very much affected by the 'quality' approach. Think about who your customers are. Think about your process and your customer's processes.

Response variables are indicators of aspects of a process that are important. In most situations several response variables will be of interest.

In thinking about response variables you should consider the following.

- Ask your immediate customers, whether internal or external to your organization, what matters to them.
- Ask your ultimate customers (the consumers of the products and services that you contribute to) what matters to them.

SUMMARIZE BELIEFS AND UNCERTAINTIES

- Be prepared to measure several response variables. There may be many aspects to the output from your process. Measurements made routinely because they are quick and cheap may not tell all of the story. And beware of the classic fallacy of giving priority to things that are easy to measure over things that are difficult to measure.
- Consider using measures of market value, cost, profit and added value. For instance, the market value of slightly mouldy grapes or slightly scabby potatoes is a more sensible response variable than is the percentage of mould or of scabbiness. A procedure for handling grapes which resulted in all bunches being about 2% mouldy would be much less commercially viable than a procedure that resulted in 2% of bunches being about 100% mouldy and the other 98% of bunches having no mould.
- Consider measures of sensitivity, reliability and robustness.
- Be prepared to use subjective assessments as well as objectively measurable quantities.
- Look at specifications, but remember that a customer's stated specifications are only a codified version of what is really required – the real requirement is that your output produces no problems when used as inputs to their process.

In some situations there are variables which might be regarded as response variables some of the time and regarded as input variables at other times. This is most likely to occur when experimenting on a series of processes, with the outputs from one process being the inputs for another process. Any measurement made on the intermediate things (which may be objects, materials, or intangibles such as decisions) can be a response variable for an experiment on an early process or a factor for an experiment on a later process.

Example: Scoring decay in timber

Pieces of timber were scored on a scale from 8 (completely sound) to 0 (completely rotten) every week for 48 weeks. Should we use the number of weeks to decay to a score of 3 or the score at the end of the 48 weeks?

The number of weeks to decay to a score of 3 is not known for pieces of timber which are sound at the end of 48 weeks, so it cannot be used as a summary response variable. The score at the end of 48 weeks has the undesirable feature that it does not differentiate well between pieces of timber which only survived exposure to the particular combination of damp soil, mould and termites for a short time and those which survived almost 48 weeks before reaching a score of 3. A reasonable composite of the two measures would be to use the score at the end of 48 weeks if this is 3 or greater and otherwise to use 6 minus 144 divided by the number of weeks to decay to a score of 3. This gives responses of -3, -2, -1, 0, 1 and 2 for pieces of timber reaching a score of 3 after 16, 18, 20.6, 24, 28.8 and 36 weeks respectively.

Example: Testing skid resistance of flooring tiles

It is desirable to be able to test the skid resistance of flooring tiles, so that manufacturers of tiles and people selecting types of tiles can make better decisions.

What to we mean by the 'skid resistance of flooring tiles'? This question becomes harder to answer the more closely we consider it.

- People walking over the tiles have a variety of types of shoes.
- People walk in different ways and in various directions over a tiled area.
- A reduction of the coefficient of sliding friction which does not obviously affect one person might lead another person to fall and break their pelvis.
- A person's shoes may be clean or dirty, new or worn, dry or damp.
- The tiles may be new or worn, recently steam-cleaned or somewhat dusty.

It is convenient to define a standardized skid resistance test, not with any expectation that such a test measures the most appropriate quantity, but to ensure that testing done at different times in different places is reasonably consistent. From a tile manufacturer's point of view, standardization of the test allows a clear statement about product quality to be made and allows legal liability to be kept under control.

One standard test method for new tiles is to prepare the surface by wiping with a damp cloth and allowing the tile to dry. The tile is then tested using a pendulum tester which drags a piece of rubber over the surface of the tile. The normal force between the rubber and the tile is kept reasonably constant by a spring. The distance over which the rubber slides is kept reasonably constant, by adjusting the height of the surface of the tile relative to the pendulum tester. The coefficient of sliding friction between the rubber slider and the tile affects the loss of kinetic energy of the swinging arm of the pendulum tester. This is assessed by measuring the height to which the arm of the pendulum tester swings.

If the type of rubber were changed, then we should consider that we were measuring a different response variable. Using a soft rubber in the pendulum tester is relevant to measuring how likely people wearing shoes with soft rubber soles are to slip. Using a hard rubber in the pendulum tester is relevant to measuring how likely people wearing shoes with hard rubber soles are to slip. I imagine that these two different response variables would be influenced by the roughness of the tiles on different scales.

3.5.2 Looking for Factors

For response surface experiments the factors are generally obvious. Here we concentrate on factor-screening situations.

The factors are the things which will be varied in an experiment. Choosing them is one of the most important aspects of designing an experiment. The choice should not be difficult if you have thought about the purpose of your experiment and understand your variables.

It is often a good idea to conduct a brainstorming session in order to produce a list of candidate factors, possibly in the form of a cause-and-effect diagram. This helps people to regard the experiment as being something that they are involved in rather than something which has been imposed from outside or above, as well as reducing the chances of finding that a possible factor has been overlooked.

3.5.3 Types of Data

Ask some of the following questions about measurements, whether they are factors or responses.

SUMMARIZE BELIEFS AND UNCERTAINTIES 51

- When do the results of measurements become available? Could this timing be changed? If so, for what cost?
- Is this variable merely a cheap proxy for other variables which are difficult, slow, expensive or impossible to measure? For instance, classification of an object as one of very poor, poor, fair, good or very good may be regarded as a cheap proxy for a more careful assessment.
- Are some quantities unknown or unknowable? Does this put a different light on the measurable quantities?
- Is this variable objectively measured or is it subjective?
- Is this variable well defined or must an operational definition be constructed? Such as 'number of teeth' needs an operational definition so that people know whether or not to count teeth which are not wholly natural and functioning normally for the various possibilities.
- What is the precision of sampling and measurement or testing?
- Would it be possible for this measurement to be improved without commensurate improvement in the things that really matter?

Ask the following questions about the variables which are being used as factors. The answers will help you to summarize what you know and what you don't know, and lead you to select a strategy for experimenting.

- Is this variable on a continuous scale (like temperature), discrete on an ordered scale (like number of children), or a discrete value from a set of unordered categories (like name of operator)?
- Is this factor a controllable factor (i.e. you wish to select the best level)?
- Do you wish to choose the best value for a variable and expect that this will remain the optimum, or will the setting be adjusted using some regularly obtained information?
- Is the variation in this factor indicative of the variation which must be expected over the use of the product being manufactured? (Taguchi uses the term 'indicative factor' to describe such factors.)
- Is this factor a block factor (uncontrollable, but helps to explain the variation between results obtained)?
- Is this factor merely a source of noise? Does it affect the results obtained in an independent way?

3.6 CONSIDER POSSIBLE TRANSFORMATIONS OF BOTH RESPONSE VARIABLES AND FACTORS

Guess which interactions between factors need to be considered. Consider theoretical knowledge, analysis of historical data and possible transformations of both response variables and factors in order to reduce the likely number of important interactions.

As was remarked at the start of the previous section, it is sensible not to worry too much about the distinction between response variables and factors, because some quantities will be response variables some of the time and factors some of the time.

Transformations are an important way in which our background knowledge can be incorporated into the design and analysis of an experimental programme. This can influence the efficiency of experimentation. It can also influence the reliability of the conclusions reached. Extrapolation of conclusions beyond the range of experimental results is much less dangerous if transformations based on accepted scientific principles or reliable models have been used than if such transformations have not been used.

3.6.1 Transformations of Response Variables

There are two good reasons and one not-very-good reason for considering a transformation of a response variable.

- To make an explanatory model for trends likely to be simpler. For instance, it is common to take logarithms of counts of micro-organisms and logarithms of chemical concentrations because the logarithms are often more simply related to temperature, time and other factors than are the counts or concentrations themselves.
- To make the variance constant (well, nearly constant) across the space of treatments which are being considered.
- People sometimes worry about normality of errors, but this is generally much less important than the other two points.

Transformations commonly used are discussed below.

Logarithm   Logarithms are often taken of chemical concentrations or counts of micro-organisms. The error standard deviation is often proportional to the response (i.e. you make errors which are 10 times as big when measuring something 10 times as big) in such situations. The logarithm has constant variance.

Logarithmic transformation is often suggested by background knowledge.

- When working with chemical reactions, it is common practice to use logarithms of concentrations of chemical constituents and the logarithm of the absolute temperature. Why? Because experience tells us that relationships are likely to be more simply expressed in such terms.
- Microbe populations increase exponentially with time, so logarithms of microbe counts often satisfy simpler models than raw counts.

If some data are zero, half the minimum detectable amount is added to all data before taking logarithms. This avoids attempting to take the logarithm of zero in the most generally accepted way.

Logarithms to the base 10 are more widely understood than natural logarithms, but natural logarithms are easier for experts to interpret because a difference of 0.01 corresponds to 1%. If results are presented graphically then the distinction should not matter because axes can be labelled using the original scale.

Reciprocal   Time to failure is often difficult to model. Factors often have a simpler effect on the rate of failure. This might suggest a reciprocal transformation. Alternatively, hardness of metal might be measured rather than the time for the wear to exceed a given tolerance.

Square root   Counts which are expected to be Poisson distributed (counts of insects, for instance) are sometimes transformed by taking square roots. This approximately standardizes the variance, but tends to imply that the effect of a treatment varies with the number of insects. Use of generalized linear models is preferred.

SUMMARIZE BELIEFS AND UNCERTAINTIES

Other powers There are times when other powers of response variables are reasonably sensible. For instance, the rate of flow of water over a V-shaped notch is proportional to the $2\frac{1}{2}$ power of the height of the water above the bottom of the V.

When looking at a quantity of particulate material we may be interested in the number of particles, the proportion by mass retained when the material is sieved, the total surface area (which influences chemical reaction rates), the mass or the volume.

Angular transformation This is also referred to as the arcsine transformation, but this is inaccurate as its mathematical formula is arcsine(sqrt(x)). It is sometimes used for analysing data which are proportions. (Note that percentage data must first be divided by 100.) The arcsine is an angle which can be measured in degrees, when its range will be 0 to 90, or in radians, when its range will be 0 to 1.571.

Like the square root transformation, this transformation is not entirely satisfactory because the effect of a treatment is unlikely to be linear on the scale of the transformed variable. Again, use of generalized linear models is preferred.

The economic significance transformation Consider attempting to translate a variable into monetary terms, if only approximately. Given data on amounts of imperfection on fruit or vegetables it is often sensible to transform these numbers so that you are analysing the economic significance of the defect. The translation from degree of imperfection to economic loss is not precise, but such a translation is sensible because presentation of the results to decision-makers is thereby simplified.

A 'gut feeling' transformation? Sometimes people say that, although some conclusion appears not to be entirely justified, they have a 'gut feeling' that the conclusion is true. If the gut feeling is nothing more than prejudice or wishful thinking, then detailed examination of its basis is unlikely to be helpful. However, sometimes a detailed examination of the basis for the gut feeling will suggest that different data should be collected or that the data should be transformed in some way.

For instance, imagine an experiment to see whether attending a training module improves the efficiency of a group of workers. Many measurements of work efficiency might be made. Somebody's gut feeling as to how the behaviours of workers have changed and why this should lead to improved efficiency might suggest a transformation of the work efficiency measurements which allows the influence of attending the training module to be discriminated from the many sources of noise in such data.

3.6.2 When Transformation for one Purpose does not Suit another Purpose

If the transformation required in order to stabilize error variance is different from the transformation required to simplify the form of the model then it may be desirable to use a generalized linear model. Seek help in these cases.

3.6.3 Transformations of Factors

Factors may be transformed in order to make models simpler. For constant predictive power, a simpler model is usually better.

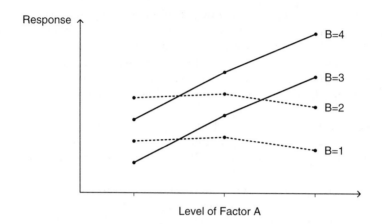

Figure 6. Illustration of situation where the effect of factor A is of one form for two values for factor B and of another form for two other values for factor B

One type of the transformation of factors is exactly the same as the transformations which can be made of response variables. Such transformations only consider one factor at a time.

We will now look at transformations which use more than one factor at a time. These may be derived using theoretical knowledge, or by the desire to avoid experimental regions which are dangerous, irrelevant or silly. These transformations can be regarded as attempting to reduce the importance of interactions. We will discuss the meaning of interactions before discussing examples of how redefining factors is likely to reduce the importance of interactions.

3.6.4 Interactions

An interaction is when the effect of one factor varies with the level of another factor (or with some combination of levels of other factors when there are several factors). To indicate interactions for qualitative variables, the effect of one factor may be tabulated or plotted for each level of the other factor. For quantitative variables it is often most convenient to use a mathematical formula involving the product of the variables.

In the abstract, it does not matter whether we say that the effect of factor A varies with the level of factor B or we say that the effect of factor B varies with the level of factor A. In some circumstances, it is important to choose how we express the interaction. For instance, Figure 6 illustrates a situation where the effect of factor A takes one form for two values for factor B and takes another form for two other values for factor B.

3.6.5 Reparametrizing Factors

If you expect substantial interactions then you might consider reparametrizing your factors. For instance, a photographic experiment would be better off using factors exposure (which is: lens aperture × time of exposure) and lens aperture rather than

SUMMARIZE BELIEFS AND UNCERTAINTIES

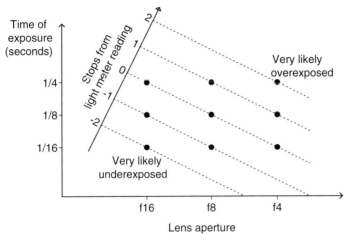

Figure 7. A first set of nine trials for a photographic experiment. Lens aperture and time of exposure are used as factors

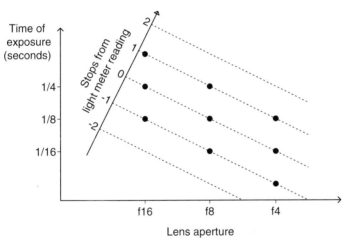

Figure 8. A second possible set of trials for a photographic experiment. Lens aperture and stops from light meter reading are used as factors

using lens aperture and time of exposure. Compare the experiments illustrated in Figures 7 and 8. In the first case, shown in Figure 7, where lens aperture and time of exposure are used as factors, interaction between the factors is likely to be very important, and a straightforward experimental design tends to result in the performing of stupid experiments (illustrated by points at the combinations of factor levels investigated) where lens aperture and time of exposure were either both large or both small, so that a photograph is likely to be either underexposed or overexposed.

In the second case where exposure (expressed as stops departure from light meter reading) and lens aperture are used as factors, a straightforward experimental design, in which three apertures are used with each three exposures, seems more sensible.

Whenever interactions are thought likely to be substantial compared to main effects, it is worth contemplating a reparametrization of the factors. Perhaps the pattern of effects would be simpler if viewed from a different perspective. If a reparametrization of the factors is very successful, then experimental data might find no interaction between the reparametrized factors – in which case you need to be very careful with the use of the word 'interaction'. For instance, the performance of the wings of a model aeroplane might find no interaction between the factors 'wing area' and 'aspect ratio'.

Reparameterization of factors is not normally done on the basis of a conjecture that it would provide an improvement. If there is doubt then it would be better to stick with the variables that are easier to understand.

Avoiding impossible, dangerous or silly combinations

Sometimes reparametrizing factors is desirable in order to avoid impossible, dangerous or silly combinations of levels of factors.

- Impossible combinations of levels of factors might be specified as 20, 40 or 60% of ingredient A, 10, 20 or 30% of ingredient B and 10, 20 or 30% of ingredient C. The fractions cannot possibly exceed 100%. This problem could be avoided by, for instance, setting the amount of C to 20, 40 or 60% of the material which is not A or B.
- A temperature of 280 °C might be interesting for one set of conditions but dangerous for a second set of conditions. Rather than using temperature as a factor, the difference from the maximum safe temperature might be used as a factor.
- An example of a silly combination of levels of factors is the use of three amounts of two different chemicals, but where one of the amounts is zero. It is generally reasonable to presume that the effect of using a zero amount of a chemical does not depend on the chemical!

Example: Use stoichiometric excess

When several factors are being varied in a chemical experiment it is likely to be a good idea to use stoichiometric excess of some chemical as a factor, rather than simply the amount of that chemical.

Example: Experimenting on hydrocyclones

A hydrocyclone is like a funnel with a lid and two extra holes. Its purpose is to separate the solid particles in a slurry on the basis of either particle size, density or a combination of these. The slurry is pumped in through an inlet tangential to the outside surface of the hydrocyclone where its diameter is greatest. The slurry can leave through either of two apertures.

One aperture corresponds to the small end of a funnel and is called the spigot. The larger and denser particles from the slurry tend to leave through this aperture. The remaining aperture is called the vortex finder. It is a tube through the lid of the funnel. The finer and lighter particles tend to be over-represented in the material leaving the hydrocyclone by this aperture. Also, the proportion of solids in this slurry is lower than in the slurry leaving by the spigot.

If theoretical knowledge were ignored then the variables which might be used as factors in an experiment are inlet diameter, spigot diameter, vortex finder diameter, flow rate of the slurry and pulp density. However, a simple factorial experiment based on these variables might suggest some ridiculous or non-feasible experimental runs: e.g. high flow rates with small diameters, or small flow rates with a large spigot so that all of the slurry simply flowed out through the spigot. Theoretical knowledge tells us that flow through an aperture varies with aperture area rather than diameter.

It might also be sensible to think in terms of the power required for pumping rather than slurry flow rate. Experimentation might be based on factors such as pump power per litre of slurry, ratio of inlet area to the sum of the areas of outlet apertures, and ratio of area of the outlets.

I would suggest that attempts at mathematical modelling should precede experimentation for this situation.

Example: Mixing paint

A colleague, Alan Veevers, told me about an experiment on paint formulation which turned out to be a failure. Some of the trial formulations did not behave like paint. They were solid!

A statistician who had just completed his Ph.D. designed the experiment. There were many factors. He consulted with several people in order to find out sensible levels for the factors. These people told him the smallest and largest commonly used amounts for the various ingredients, because he had asked for that information. However, they did not tell him that some of these ingredients were alternatives which should not be varied independently and that the ratio of liquid to solid ingredients should be kept within reasonable bounds. He hadn't asked some questions which, with hindsight, he should have asked.

Example: Making a cake

Recipes for cakes often say that the final liquid ingredient (usually milk) is to be added until the mixture reaches a particular consistency. Due to the variable size of eggs and the variable moisture content of flour this way of specifying the amount of liquid is likely to give more repeatable results than would specifying the absolute amount of liquid.

Example: Plasma etching

Following an experiment on plasma-etching of an aluminium-silicon layer, a substantial two-factor interaction was found. These were the total pressure in the reaction chamber in which the plasma-etching took place and the flow rate of chlorine.

A single factor, partial pressure of chlorine, was found to provide as much explanatory power as a model with two factors for the rate of etching. This alternative way of parametrizing the factors accords with chemical theory for reaction rates. See Robinson (1993a) and the references therein for further details.

3.7 DOCUMENT YOUR SUMMARY OF BELIEFS AND UNCERTAINTIES

The beliefs and uncertainties of the people whom you have consulted should be summarized in some manner. This includes beliefs and uncertainties about experimental protocol, likely sampling and testing variability, and the effects of factors on the response variables of interest. Where people have different beliefs it may be desirable to resolve some of the differences as early as possible in order to promote team harmony.

Questions are more important than answers. I have met many researchers who, in principle, would agree that questions are more important than answers, but who seem to think that objectivity is even more important. I disagree about the importance of objectivity. I believe that we can only do research efficiently by allowing our subjective opinions and our value judgements to influence our choice of questions. However, we should then endeavour to ensure that our efforts to answer those questions are not contaminated by our subjective opinions and our value judgements.

3.7.1 Beliefs about Experimental Protocol

You should document your intended experimental protocol and any alternatives that you wish to consider. If there are any issues of getting permissions, authority, personal responsibility, safety, ethics or possible environmental damage then these should be discussed.

- Where will raw materials be obtained?
- What sampling methods will be used to obtain portions of raw materials?
- How will trials be set up and conducted? How will you know when to terminate the trials?
- How will response variables be measured? In detail, what sampling and testing methods will be used?
- How will variables be kept constant should this be required? How precisely can they be set or controlled?
- Perhaps some nuisance factors cannot be controlled. Can they be measured or indirectly monitored? How stable are they likely to be during the period required for experimentation?
- How will the levels of factors be altered? What are your beliefs about the expense of altering hard-to-adjust variables?
- Roughly what total budget (time, money, personnel, access to sites, access to specialized equipment, etc.) do you have for experimenting? How might this budget change?
- What precision is expected?

3.7.2 Construct an 'Influence Matrix' for Main Effects

The idea of an 'influence matrix' may be familiar to some readers who are aware of Quality Function Deployment. It is useful for factor-screening experiments.

SUMMARIZE BELIEFS AND UNCERTAINTIES

It consists of setting up a rectangular table of response variables by factors and indicating the likelihood that a factor influences a response variable, in the opinions of the people who have been consulted. Any agreed symbols or scale can be used, such as the following:

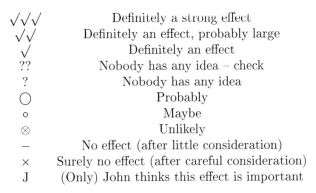

Example: Wave soldering machine

For an experiment on a wave soldering machine as discussed on page 81, an influence matrix like Table 1 might be drawn up. Interested people should reach agreement that all entries are fair summaries of the state of beliefs and uncertainties.

3.7.3 Beliefs about Variability

You should document estimates of variability which are relevant to your proposed experimentation, together with an indication of the relevance of these estimates.

Table 1. Influence matrix summarizing opinions about a wave soldering machine

FACTORS	RESPONSES				
	Solder splashes	Blow holes	Holes not soldered	Legs not soldered	Uniformity
Solder pre-heat temp.	⊗	–	○	○	?
Solder temp.	○	○	○	○	?
Conveyor speed	√	○	○	○	??
Time in solder wave	⊗	○	○	○	○
Omega wave freq.	?	–	○	○	○
Wave height	√	○	√√√	√√√	√
Flux density	○	○	√	√	○
Type of board	○	√	√	√	√
Air knife pressure	×	√√	?	?	?
Flux contamination	×	?	○	○	○
Solder pot clean?	?	?	○	○	○
Solder oil?	⊗	×	?	J	⊗

- Do you know the error variance from similar experiments conducted in the past? Were those experiments similar enough to the proposed experiments for those estimates of error variance to be regarded as useful predictions of the error variance likely in the proposed experiments?
- How precise are the measurement techniques? Is past precision a good predictor of future precision?
- When measuring equipment is calibrated, how large are the adjustments which are typically made?
- How large have the sampling errors been when working with materials of the same type? Was the moisture content and particle size similar to what is expected in the proposed experiment?
- Is variation between operators or between machines often a problem for experimental work of the intended type?
- How routine will the proposed work be for the people who will be required to do it?

4

Decide on a Strategy

This chapter aims to help you to decide on your broad experimental strategy. One strategy – somewhat extreme – would be to conduct a single, all-embracing, super-duper experiment. Another extreme strategy would be to conduct trials one at a time, carefully considering what has been learnt and what further information needs to be obtained after each result becomes available.

Generally, it is most practical to think of experimentation as a sequential, staged process, consisting of a number of simple experiments. You might, for instance, plan to spend 10% of your resources on a preliminary trial to find a sensible range for two factors, to spend 50% of your resources on a substantial trial involving a large number of factors, and to keep 40% of your resources in reserve to be used as indicated by the results available at that time.

Strategies tend to vary according to the type of experimental objective. General principles behind the choice of strategy will be discussed first. Then these principles will be applied to the various types of experimental objective and to some specific examples.

Costs and benefits of experimenting

One useful way of comparing possible experimental strategies is to compare the costs of experimenting and the likely benefits of the information obtained by following the possible strategies.

The value of the information likely to be obtained by experimenting can be estimated by thinking about possible improvements to processes and the likelihood of achieving them. The present value of this information is often strongly influenced by the time delay in waiting until experimental results become available.

The costs of experimentation are easier to estimate than is the likely benefit of the information. These costs include the time of people to plan the experiments, conduct the experimental runs, analyse the results and report the conclusions. They also include the direct costs of conducting the trials. They may include costs due to interference with normal production or costs associated with disposal of unsalable products during trials.

The first questions that should be asked once you have an approximate idea of the costs and benefits of experimenting are 'How many runs can you afford to make?' and 'Is experimentation justified at all?' Some additional remarks are as follows.

- For agricultural and horticultural trials, it is often only possible to run one set of experiments per year, because many plants only yield once a year. A experiment with a complicated structure and a large number of runs will be finished in about the same length of time as a small, simple experiment. In contrast, many industrial experiments can be completed in an hour, a day or a week, with a subsequent experiment being started quickly after completion of previous experiments. A single large experiment with, say, 30 runs will not be much cheaper to conduct than a series of smaller experiments with the same total number of runs, say three experiments with 9, 9, and 12 runs.

  Some books about experimental designs deal with complicated theory for large and complicated designs. Complicated designs are most likely to be appropriate when the cost of experimenting is dominated by the number of stages in your experimental strategy, as for agricultural trials. Simpler designs are most likely to be appropriate when the cost of experimenting is dominated by the number of runs performed.
- The potential benefits of medical trials include reducing the risks that treatments being tested might have undesirable side-effects under certain conditions or for some classes of people. The costs of medical trials on human subjects include the opportunity cost of failing to treat the subjects in another way which may have been better.
- The benefit of getting information may or may not be sensitive to the precision of the estimated effects of factors being investigated. For instance, suppose a new drug for treating malaria produced nausea in 1% of a sample of 200 people. This would not be a barrier to further investigation and development even though the precision of the estimated '1%' is very poor. In contrast, suppose that a current metallurgical process extracts an average of 73.4% of the gold for run-of-mine ore in a particular deposit. An experiment that found that an alternative metallurgical process would extract an estimated 75% of the gold would be of much greater interest if the standard error of the 1.6% improvement was 0.6% than if the standard error of the 1.6% improvement was 3%.

  In general, low-precision results are useful for exploratory work and for development of new processes with no strong competitors, while high precision is required for making small changes to well-known processes in situations where there are many competitors.
- The last phase of experimenting is often a 'confirmation' trial in which the apparently best combination of levels of factors is checked to see if it really is as good as it appears to be from earlier experiments. The decision as to whether or not to conduct such a trial should be based on a careful estimate of the benefits of avoiding some risks, and the costs associated with experimenting and with delaying implementation. I suspect that in most circumstances, even for simple investigations, it is a good idea to perform a 'confirmation' run. Often, this will take the form of implementing the apparently best procedure and using it in a cautious mode, with more-than-usual effort being expended to monitor the results.

Uncertainty about experimental technique

One issue that can affect your choice of experimental strategy is that you may be worried that your experimental technique is not completely reliable. If you are not

DECIDE ON A STRATEGY

confident of your experimental technique then it is always a good idea to adopt a sequential approach.

Example: Comparison of SCRIM vehicles

It was intended to compare two vehicles used for testing the skid resistance of road surfaces. These vehicles are called Sideways Force Coefficient Routing Investigation Machines, and the acronym SCRIM is commonly used. In essence, they are small trucks which have a wheel underneath the middle of them as well as the usual wheels near to the four corners.

This central wheel is mounted so that it points in a slightly different direction from the other wheels. The vertical component of the contact force between this wheel and the road over which the SCRIM is travelling is kept as constant as possible. The horizontal component of force square to the direction of rolling of the wheel is monitored with a force transducer. This sideways force is used to estimate the coefficient of sliding friction for the contact between the wheel and the road.

The tyre on the central wheel is changed frequently and is kept very wet during the operation of a SCRIM. The drivers of SCRIMs endeavour to position the central wheel over the wheel-paths which have been most worn by traffic, so that the measured coefficient of friction indicates whether the road needs to be resurfaced.

I was asked to advise on an intended experiment comparing the SCRIM machine in New South Wales with the SCRIM machine in Victoria. The intended experimental protocol was to have the two machines follow one another, testing the same segments of road. It was thought that there might be differences due to mechanical differences between the vehicles because of slight variations in design and maintenance practices, in the state of tyres on the central wheel, in water flow rates, in where the SCRIMs are positioned relative to the most worn wheel-paths, in driving speed, in calibration of force transducers, in temperature and in the state of the road surface caused by passage of previous SCRIMs.

The New South Wales SCRIM was only intended to be in Victoria for about two weeks, so a pilot experiment was run using only the Victorian SCRIM. The factors which could be investigated using only one SCRIM were investigated. Tyres were swapped. Different drivers were used. Transducer calibration was repeated. The water flow rate was varied. First and second passages of the SCRIM over the same route were compared.

It turned out that the main benefit of the pilot experiment was to highlight problems with the experimental protocol which had not been anticipated. It was much better to have found out about these problems during a relatively less expensive pilot experiment than to have only discovered the problems during a more critical and more expensive experiment.

Uncertainty about components of variability

Another type of uncertainty that may influence your choice of experimental strategy is that you may be uncertain about the variability associated with experimental runs. One reason for adopting a sequential approach is that it would be easiest to design an experiment after it was finished and completely analysed. At that time it is easier to decide on such things as:

- the number of experimental runs;
- which factors to include;
- what levels to use for factors, and;
- what precautions ought to be taken to avoid disasters.

Uncertainty about trends and about the location of the region of interest

Possibly the most common type of uncertainty which affects the choice of experimental strategy is that you are uncertain how the factors of interest influence the response variables. At the start of experimenting, you do not even know which region of possible levels for the factors is of greatest interest. You will be most interested in the the region near the best set of levels for the factors, but you don't yet know where this is.

If you were only intending to perform a single experiment then you would be likely to be very cautious due to fears that you might use poor experimental techniques, waste resources investigating factors and levels which turn out not to be interesting, or cause some type of disaster. For instance, before the first testing of a hydrogen bomb there was some concern that a chain reaction in ordinary water might completely obliterate life on our planet!

It is better to start with a small, poorly planned experiment and to get some information and some experience which helps with the planning of the next stage of experimentation, than to hope that a single experiment will turn out well.

Often we do not know where is the most interesting region until a few trials have been done. The need to home in on the region of interest is a major reason for using a sequential approach to experimenting.

Near an optimum, it is expected that the main effects will be small. This may mean that interactions and quadratic components of the main effects are larger than the linear components of the main effects.

Some useful general principles for deciding on the structure of a sequence of experiments within a large study are the following. They are all related to the fact that experiments early in a sequence will often provide information which helps in the planning of later experiments.

- Don't try to achieve everything in a single experiment. Something will almost always go wrong. One formulation of this principle is the so-called 25% rule:

    Don't spend more than 25% of your resources on your first experiment!

  I don't suggest that you adhere rigidly to this rule, but when violating it you ought to be able to explain why you have done so in terms of information from other sources.
- Study your sampling and measurement system reasonably early.
- Think of some experiments as being exploratory and some as being confirmatory.
- Check that important factors have not been forgotten before studying the effects of other factors in minute detail.
- Be prepared to change strategy after seeing the results, particularly the results of early experiments.

DECIDE ON A STRATEGY

A friend, Neil Fothergill, suggested that a sequential approach to experimentation can be regarded as a successive approximation technique. We would like a mathematical function which tells us approximately how a response varies with the experimental factors over a range of interest. Both the region of interest and the mathematical function are refined by successive experiments, but there would be little point to getting a very accurate approximation over a range which was not of interest.

Generic chemical example of a sequence of experiments

Experimental studies typically involve several phases. For instance, consider a chemical process which we would like to improve. For simplicity, assume that only one aspect of the process needs to be improved, say the yield. A number of factors might affect the process:

- purity of raw materials: commercial or laboratory grade;
- parameters of recipe used;
- catalyst;
- temperature;
- pressure;
- time of reaction;
- type of reaction vessel;
- agitation rate.

The basic steps in optimizing the process are

1. Check that sampling and testing procedures are in order. Check other aspects of experimental method also.
2. Find out which factors are important. This is often referred to as 'factor screening'.
3. Find out approximately what values for the important factors will give the optimum.
4. Find the optimum to appropriate precision, using an experimental region near to the optimum.
5. Check that the proposed model provides adequate predictive power.

In any real situation, some prior information about the effects of these factors will be available. There are likely to be some restrictions on the levels available for some factors, due to restrictions on equipment, safety regulations and availability of materials.

If we know very little about the process, we might start by doing a fractional factorial design which investigated all of the factors at two levels only in a minimum number of runs. This would be sensible if experimentation were expensive and all factors can be altered with equal ease or difficulty. If some factors are more difficult to alter than others, then this factor screening part of the sequential procedure might investigate some factors separately from others. For instance, it might be convenient to try out most factors in an experiment involving only one type of reaction vessel and one catalyst; and to investigate the effects of reaction vessel and catalyst at a later date.

Next, we might try some small 2^n factorial experiments or fractions of 2^n experiments in order to determine the extent of the region of interest. Then, some 3^n factorial experiments or fractions might be used to find the optimum more precisely and to estimate the response surface in the region of the optimum so that the criticality of each of the factors can be assessed.

Generic drug testing example

There are many stages to the testing of new drugs. A strategy for experimenting in a large number of stages is appropriate both because of the structure of the costs and benefits of experimenting, and because of the large number of uncertainties which must be addressed before a drug can be made available for medical use.

- A proposed new drug might not be effective.
- Its effectiveness might vary between animal species, so trials on, say, rats cannot reliably be extrapolated to humans. Even if experiments have been done on several non-human species including species thought likely to be similar to humans so far as a particular drug is concerned, extrapolation is still unreliable.
- A proposed drug might be toxic in some circumstances or in combination with other drugs or treatments.
- The dose required in order to achieve the required effect is not known. Does it vary with the age, gender or weight of the person taking the drug?
- The maximum dose which must not be exceeded in order to avoid undesirable short-term effects is not known. Does it vary with the age, gender or weight of the person taking the drug?
- Are there problems associated with long-term use of the drug (both on individuals taking the drug and on the development of drug-resistant strains of bacteria or similar environmental effects)?
- Might there be ways of reducing any undesirable side-effects?
- The benefits of drug testing arise because a very small fraction of the drugs tested turn out to be very useful. I'll refer to them as 'winners'. The design of strategies for testing drugs aims to stop experimenting on drugs which are not going to be winners as soon as possible in order to reduce costs. However, because the winners are very valuable, drugs are not dropped from experimental programmes unless there is clear evidence that they are not winners.
- Some possible experiments, particularly large trials on human subjects, are very expensive (in monetary terms as well as in terms of human life and human misery). These trials are not permitted unless there is evidence that the risk of disaster is not too great. Some experiments are done as screening trials in order to reduce the risk of other experiments.
- For drugs which turn out to be winners, it is best (both for society and for the cash flow of the drug companies) that the experimental programme be completed as quickly as possible.

The point of this example is not to provide advice about drug-testing. Other people have much more specific knowledge about that topic than I do. The point is to illustrate that when there are many components to your uncertainty it is likely that there will be many stages in a sensible strategy for experimenting. A crucial decision that must be made in such circumstances is the order to tackling the uncertainties.

Example: Relays for telephone switchboards

One of the most economically significant experimental efforts on a large number of factors was the development of wire-spring relays for use in crossbar switchboards. These were developed in the 1950s. Taguchi (1987, pages 118–9) explains that the effects of at least 2000 factors on several tens of characteristics were studied in Japan. At this time, Taguchi was a lecturer on experimental design at the Electrical Communication Laboratory of the Japan Telephone and Telegraph Co. which coordinated research done by manufacturers in related industries.

On the insulating material alone, experiments were conducted considering types of paper, types of varnish, abrasion-resistant additives and processing methods. The final result was a relay which was longer-lasting (it would work over 600 million times without malfunctioning) than a competing product developed at Bell Laboratories in the United States, and which had wider specifications for its components. Taguchi attributes the success of this R&D experimentation to the large number of factors investigated. He gives the cost of the research as 2000 million yen and suggests 'the social profit and profit by export which were obtained must have exceeded 1000 billion yen'.

Example: Operation of atomic absorption spectrometers

Cellier and Stace (1968) describe two experiments which established the essential operating characteristics of atomic absorption spectrophotometers. Much previous study by other investigators varying one factor at a time had not been as satisfactory.

The first of these experiments was half of a 2^7 factorial experiment plus 12 centre points. The second was a central composite design in the four factors for which substantial interactions had been identified in the first experiment. The first experiment had 76 runs and the second had 31 runs.

4.1 MAIN EFFECTS AND INTERACTIONS AND THEIR RANGES OF VALIDITY

Before discussing experimental strategies, it is necessary to introduce a few concepts.

The complexity of a model for the trend depends on how widely valid you need it to be. It is related to how well mathematical functions can approximate reality over a given range. For instance, we need more terms of the Taylor series $1+x+\frac{1}{2}x^2+\frac{1}{6}x^3+\ldots$ for e^x in order to have specified accuracy over a wider range. Two terms, $1+x$, give an approximation within 10% over the range from -0.39 to 0.53. Three terms give an approximation within 10% over the range from -0.70 to 1.10. Four terms give an approximation within 10% over the range from -1.01 to 1.74. Five terms give an approximation within 10% over the range from -1.31 to 2.43.

When experimenting sequentially it is generally important to find out the interesting range as quickly as possible. It is only valuable to get detailed knowledge once you become confident that you are looking at the interesting range.

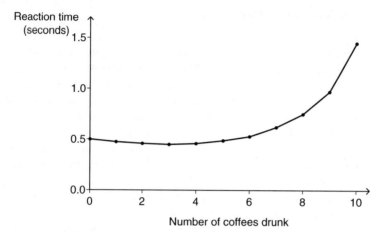

Figure 1. Effect of number of coffees on reaction time (for no whiskies)

4.1.1 Degrees of Complexity of Models

It is often true that the range over which a factor is investigated affects which type of model is appropriate. This will be illustrated first by an example, then discussed in the abstract.

Example: Coffees and whiskies

Suppose that a person consumes a number of coffees and a number of whiskies over a short period of time. The person's reaction time is measured under various circumstances. The results are illustrated in Figures 1 and 2. This example is based on a presentation by a colleague, Allan Adolphson. The results presented are mythical. A real experiment would need to consider the order and spacing of the trials very carefully, and would probably show a great deal of variability in the results.

The relationship between the number of coffees and reaction time is nearly linear over the range from zero to three coffees. Over the range from zero to five or six coffees, a quadratic model would provide a better fit. The departure from a linear or quadratic model is quite important if the model needs to be valid for the range up to 10 coffees shown on Figure 1.

Figure 2 shows that, for zero whiskies up to two whiskies, the effect of the number of whiskies is unaffected by whether no coffees or three coffees have been drunk. The number of coffees does affect the reaction time (the reaction time is slightly quicker if three coffees have been drunk), but curves on Figure 2 for zero and three coffees are approximately the same shape over the range from zero whiskies up to two whiskies. A model to explain reaction time for the range from zero whiskies up to two whiskies does not need to include any interaction terms.

If we wish to make statements which are valid for the range from zero to four whiskies, however, then a model with no interaction between the factors is not

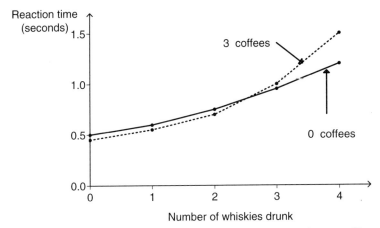

Figure 2. Effect of number of whiskies on reaction time for no coffees and for three coffees

adequate. We must now say that there is an interaction between the effects of whisky and coffee *over the range investigated*. Another way of saying that there is an interaction over this range is that the effect of whiskies depends on how many coffees have been drunk, and vice versa.

The issue of transforming response variables has been discussed elsewhere. Here it would be reasonable to consider using the logarithm of reaction time as a response variable. A logarithmic transformation would consider any 10% difference in reaction time to be equally important.

Linear and quadratic terms

Often there is not a simple linear relationship between the response variable and a factor. Provided that an experiment has at least three levels of the factor it may be possible to identify a non linear relationship.

- A linear model is likely to be an adequate fit over a narrow range which is far from the optimum.
- Near to an optimum a quadratic (or higher order) model is likely to be needed because the linear terms are always small near to an optimum.
- For a model to fit over a wide range it is common to need to choose a quadratic (or higher order) model.
- High-order polynomials tend to be unreliable near to the edge of the region of the data.

Need for interaction terms

Statements about interactions between factors are sometimes made in a rather abstract way. I believe that it is only useful to discuss the need for interaction terms relative to a specified region of interest.

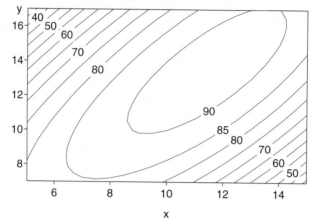

Figure 3. Contours of a function of two variables over a relatively wide region

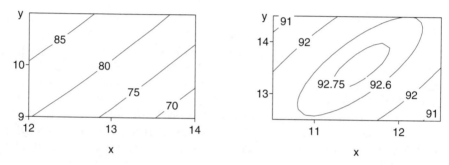

Figure 4. Contours of a function of two variables over two relatively narrow regions

That the need for interaction terms can vary with the region of interest is illustrated in Figures 3 and 4. They show contours of the same mathematical function of two variables, x and y. Figure 3 shows the function over a relatively wide range. There are a few ways of saying that there is an interaction between the effects of x and y over the range shown in Figure 3.

- The function could not be well approximated by the sum of a function of x and a function of y over the range shown in Figure 3.
- The effect of x varies with the value of y over the range shown in Figure 3.
- The effect of y varies with the value of x over the range shown in Figure 3.

In contrast, Figure 4 shows the function over two relatively narrow ranges. Over the range $12 < x < 14$ and $9 < y < 11$, the function could be usefully approximated by the sum of an effect due to x and an effect due to y. We do not need an interaction term. The range $10.5 < x < 12.5$ and $12.5 < y < 14.5$ is likely to be of more interest, since it includes a local maximum for the function. Over this range the function could not usefully be approximated by the sum of an effect due to x and an effect due to y. We do need an interaction term.

DECIDE ON A STRATEGY

It is worth noticing that over the range $10.5 < x < 12.5$ and $12.5 < y < 14.5$ the function could be well approximated without using interaction terms, if new variables were defined using, say, the directions of the major and minor axes of the approximately elliptical contour where the function takes value 92.75. Such redefinition of axes may be useful in some contexts.

Major point

The point which needs to be understood from this section is that the required complexity of a model depends on the range of the independent variables. You need to talk about the region of interest before talking about interactions or any other part of the mathematical model being used to describe trends.

- Linear models are likely to be accurate only over a small range.
- Interactions often become more noticeable as the ranges of factor levels are increased.
- Interactions and quadratic terms which seemed important in an experiment over a wide range might become unimportant if the range of the factors is decreased.

From a purely mathematical point of view it is clear that a smooth function can be approximated by polynomials. The approximation becomes more accurate as the order of the approximating polynomial is increased and becomes more accurate as the region over which the approximation is required becomes smaller. For a polynomial of degree n, provided that the $(n+1)$st derivative of the function exists, the error made by polynomial approximations decreases like a function of order w^{-n-1} as the width, w, of the region decreases. Regrettably, this mathematically reliable result is not usually what we need to know.

When fitting models to experimental data, I would offer the following suggestions.

- When all variables are continuous it is generally sensible either to fit a linear model or a quadratic model. It is seldom worthwhile fitting a higher order model, particularly when there are several variables.
- For a given accuracy of prediction, a quadratic model will be adequate over a range two to three times as wide as a linear model.
- Interpolation is much more reliable than extrapolation.

4.2 BEWARE OF ONE-FACTOR-AT-A-TIME EXPERIMENTATION

One of the most important points about choosing an experimental strategy is a negative one. Many scientists and engineers experiment by varying one factor in each trial. The purpose of this section is to argue that this is seldom a sensible strategy.

Some research workers seem to believe that Scientific Method requires that only one factor be varied at a time. For instance, Gould (1996, page 367) wrote

> Reduction of confusing variables is the primary desideratum in all experiments. We bring all the buzzing and blooming confusion of the external world into our laboratories and, holding all else constant in our artificial simplicity, try to vary just one potential factor at a time.

I agree with Fisher's (1926, page 511) comment on the suggestion that we must ask Nature one question at a time

> ... this view is wholly mistaken. Nature ... will best respond to a logical and carefully thought out questionnaire; indeed, if we ask her a single question, she will often refuse to answer until some other topic has been discussed.

I believe that attempting to define Scientific Method to everyone's satisfaction would be a waste of effort. Wallis and Roberts (1956, page 5) put this view more strongly:

> There is no such thing as *the* scientific method. That is, there are no procedures, formal or informal, which tell a scientist how to start, what to do next, or what conclusions to reach.

We may be very interested in knowing what would happen if we varied one factor at a time but, even if that is our primary focus, an experiment in which only one factor is varied may not be the most efficient way of proceeding.

4.2.1 Exploratory Trials

Isolated one-factor experiments are an excellent form of experimentation for solving simple problems. For instance, suppose it is thought that a problem with a chemical reaction might be fixed by adjusting the temperature or, less probably, the amount of a specific ingredient. Then a sensible strategy is to first try a one-factor experiment in which the temperature is varied. It is very likely that the results from such an experiment will enable the problem to be fixed. If there is still a problem then a sensible second one-factor-at-a-time experiment would be to vary the amount of the specific ingredient.

The advantages of this approach over doing one experiment in which both factors are varied are as follows.

- The expected amount of experimental effort is generally smaller. For experiments using three levels of factors, a two-factor factorial experiment would require nine runs. The one-factor-at-a-time approach is likely to require only three or six runs.
- If the levels for factors are chosen badly then less experimental effort is wasted.

However, one-factor-at-a-time experimentation is not recommended for solving complex problems.

The idea that 'we'll do a statistically designed experiment once we find out what is happening' is often the view of people who experiment with one factor at a time. It is justified only when two conditions are satisfied, and not always then.

- The amount of noise in the results obtained must be small, say, the difference between runs generally exceeds three times the standard deviation of the variation between repeats. When many replications of each treatment would be required for a one-factor-at-a-time experiment, then it is virtually always true that varying one factor at a time is inefficient.
- Results must be quickly available, so that gains in flexibility can be obtained.

A better strategy is to use small two-factor and three-factor factorial experiments as exploratory trials.

DECIDE ON A STRATEGY

4.2.2 When Experimental Runs have Poor Precision

Suppose that there are no interactions. Generally, one-factor-at-a-time experimentation will work well in such a situation. However, if there is a large amount of noise then one-factor-at-a time experimentation is generally inefficient compared to factorial or fractional factorial experiments. Czitrom (1999) makes this point very well.

For instance a 3^2 design takes the same amount of effort as three replicates of a one-factor-at-a-time experiment with three levels for the factor. The 3^2 design has a slight advantage in terms of robustness or providing reassurance that there really is no interaction.

But what if you were going to do more than one one-factor-at-a-time experiment? Consider comparisons between a 3^2 design and 10 points in the orientation of a '+' sign. The 3^2 gives equal precision for given measurement error variance.

X	X	X
X	X	X
X	X	X

	XX	
XX	XX	XX
	XX	

Suppose each single measurement has variance σ^2. For the design on the left, a 3^2 factorial design with 9 runs, comparisons of one level of a factor to one of the other levels involve comparing two averages of three runs so they will have a variance of $(\frac{1}{3} + \frac{1}{3})\sigma^2 = \frac{2}{3}\sigma^2$.

For the design on the right, two one-factor-at-a-time experiments with a total of 10 runs, comparisons of the mid-level of a factor to one of the other levels involve comparing an average of two runs with an average of six runs so they will have a variance of $(\frac{1}{2} + \frac{1}{6})\sigma^2 = \frac{2}{3}\sigma^2$. Comparisons of the two extreme levels of a factor involve comparing two averages of two runs so they will have a variance of $(\frac{1}{2} + \frac{1}{2})\sigma^2 = \sigma^2$.

Similarly, a 2^3 is better than three replicated one-factor-at-a-time experiments. Half of a 2^4 design gives the same precision as 20 runs with 4 at the origin and 4 at each one-factor alternative. And half of a 2^7 design gives the same precision as 32 runs with 4 at the origin and 4 at each of the 7 one-factor alternatives.

The general conclusion, even if there are no interactions, is that if you need to replicate to get enough precision with a series of one-factor-at-a-time experiments, then there are better alternatives. Using factorial or fractional factorial experiments will enable you to get more information for your efforts.

However, if there are any interactions, factorial or fractional factorial experiments will put you way ahead! A one-factor-at-a-time experiment is likely to need to be repeated at the levels of other factors at which you subsequently find you are most interested.

4.2.3 When there are Interactions

One of the problems with varying one factor at a time is that if there are any significant interactions between factors, then some of the one-factor-at-a-time experiments need to be repeated at different levels of the other factors. This may not be a very important problem when optimizing a process since the shape of the response surface is only of interest in the neighbourhood of the optimum.

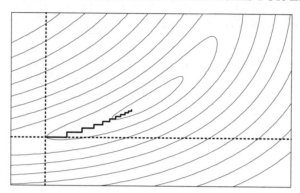

Figure 5. Contours of a response surface which has a curved ridge and a peak near the centre of the diagram

The main effect of a factor is defined to be its average effect over the range of a set of experimental runs – this is generally different from the effect of that factor in a particular region. When the effect of a factor is fairly consistent over the ranges used for all of the other factors then one may feel confident about the effect of that factor in similar circumstances.

Two-factor illustration

Consider the response surface illustrated by the contour diagram shown in Figure 5. Taking the two factors to be the two dimensions of the piece of paper, there is a substantial interaction between the factors. The effect of one factor varies with the level of the other factor. A series of one-factor-at-a-time experiments starting at the intersection of the two dashed lines would lead to the path shown as a solid line. This is north-south and east-west bushwalking. It is silly.

Furthermore, if a one-factor-at-a-time experiment has been done at some time in the past then the process is likely to be operating near a point on the ridge, so further one-factor-at-a-time experiments are unlikely to find improvements.

4.2.4 Types of One-Factor Experiments

Daniel (1973) has distinguished between five types of plans for experiments in which one variable is varied at a time.

1. 'Strict' one-at-a-time plans vary one factor from the condition of the preceding trial. These are the types of experiments that I have been discussing in this section.
2. 'Standard' one-at-a-time plans vary one factor from the standard condition. These can be of little value if there are substantial interactions between factors and the standard conditions are not nearly optimal.
   Figure 6 shows contours of two response surfaces which illustrate this point. The continuous contours show a response surface for which there is no interaction between the factors. It can be expressed as the total of an effect due to the first

DECIDE ON A STRATEGY

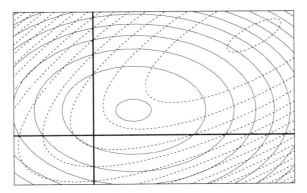

Figure 6. Two response surfaces which cannot be differentiated using standard one-factor-at-a-time experiments. The contours of one function are shown as continuous lines. The contours of the second function are shown as dotted lines. The two functions are equal all along the solid lines

factor and an effect due to the second factor. The dotted contour lines show a second response surface for which there is an interaction.

Note that the surfaces are identical on the solid lines which correspond to standard conditions.

3. 'Free' one-at-a-time plans are where new runs can be made under any set of conditions, after study of all earlier runs.
4. 'Paired' one-at-a-time plans produce two observations and hence one simple comparison at a time.
5. 'Nested' or 'curved' one-at-a-time plans are where runs are produced in sets including all combinations for one easy-to-vary factor. To my mind, these are not one-factor experiments.

If you are tempted to experiment on a one-at-a-time basis, ask yourself which of these types of plan you are following and why.

4.3 STRATEGIES FOR STUDYING SAMPLING AND TESTING ERRORS

This section deals with strategies for estimating the various components of sampling and testing variation.

It is important to study sampling and testing errors so that you know which aspects of sampling and testing are having most effect on the uncertainty of your measurements. For instance, the estimated monetary value of a shipment of coal is affected by errors in the assessment of mass of the shipment, errors in the assessment of average moisture content and errors in the assessment of other grade characteristics such as energy content, ash content and sulphur content. If one source of error dominates the uncertainty in the estimated value then it is likely to be worthwhile to reduce that source of error.

The bias of a measurement can often be reduced by frequent testing of a standard. However, this increases the total amount of testing required and increases the variance of the reported results because the results of tests on standards must be subtracted from the tests on material of unknown properties. It is necessary to have reliable information about sampling and testing errors in order to make informed decisions about whether to make frequent comparisons to a standard.

4.3.1 Strategies for Estimating One Component of Variation

In essence, the way to estimate the standard deviation of a single component of variation is to take replicates. The differences between the replicates are affected only by the single source of variation and can be used to estimate the standard deviation.

Sometimes it is reasonable to assume that the component of variation is of a roughly constant size. At other times, if you are concerned that the amount of variation might not be constant but might vary according to the average or in some other way then it is necessary to take several replicates under each of several conditions so that the standard deviation can be estimated for each of those conditions.

If the people making measurements know that the precision of their measurements is being studied then they may be particularly careful to be predisposed to making replicates look as similar as possible. Such behaviour would result in an optimistic view of the precision of the measurement procedure. This problem can be circumvented by asking people to measure the replicates on different occasions and labelling them in a way which does not allow them to be distinguished from routine samples for testing.

Example: Sugar cane sampling error

An experiment was done to estimate the variation between samples of hammer-milled sugar cane. A quantity of hammer-milled sugar cane was divided into two subsamples, each subsample was split into two, and a third sample division step was also performed, resulting in eight samples which were then tested. This procedure was repeated a few times.

The person conducting the experiments knew that the purpose of the experiment was to estimate the variability and was anxious that 'good' results would be obtained, so he removed several large pieces of the waxy part of the stalk of the cane. The remaining pieces were smaller and more uniform in structure and composition.

The results of this experiment were to be used for designing a new sampling system. Luckily, the designers of the new sampling system found out that the estimate of variability was not a realistic estimate of the variability which could be expected in the future. Otherwise, the sampling system might have been designed to take samples which were much smaller than required to achieve the desired precision.

Example: Testing bitumen

Suppose that twelve replications of a penetration test for paving bitumens on each of four different bitumens gave the following results. This data is taken from that for laboratory 4 in Dickinson and Robinson (1977).

DECIDE ON A STRATEGY

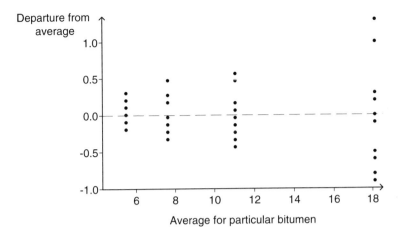

Figure 7. Plot of deviations from average against the average, for a penetration test on paving bitumens

17.3	18.3	19.4	17.5	18.0	18.3	17.2	17.6	18.4	18.0	18.1	19.1
10.6	11.1	11.5	10.9	11.0	11.2	10.7	10.9	11.1	10.8	11.0	11.6
7.6	7.8	7.8	7.3	7.6	7.5	7.4	7.9	7.5	7.4	7.6	8.1
5.5	5.5	5.5	5.4	5.3	5.4	5.4	5.4	5.7	5.4	5.6	5.8

Figure 7 shows this data in a way which is relevant to considering the variability of the test result. It shows the deviations from the average test result for the particular bitumen on the vertical axis. The horizontal axis shows the average test result, so there are columns of results for each of the four bitumens.

The only problem with Figure 7 is that it does not allow us to see where points are plotted on top of one another. This is fairly common because of the discreteness of the data. The penetration is measured to the nearest tenth of a millimetre. This problem can be overcome by what is called 'jittering', which is adding some noise to the locations of the points so that they are less likely to overlap.

In Figure 8, the points plotted are the same as in Figure 7 except that they have been jittered in both directions by an amount which is uniformly distributed between −0.05 and 0.05. It is easier to get a realistic idea of the total number of points and of their spread from the jittered graph. Note that jittering in one direction would have been adequate to achieve this.

4.3.2 Strategies for Estimating Two or More Components of Variation

The most common type of experiment for estimating the standard deviations of two or more components of variation is a hierarchical design. A simple case of this is an interlaboratory comparison in which materials are tested in each of several laboratories, with each laboratory replicating their testing. Because of the problem that such samples may be treated with special care, so that the variability observed is less than usual, it is a good idea to distribute similar samples for testing, rather than asking that duplicate samples be tested.

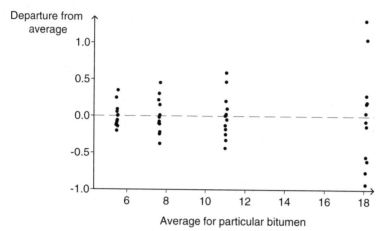

Figure 8. Jittered plot of deviations from average against the average, for a penetration test on paving bitumens. The amounts of jittering in both directions are uniformly distributed between −0.05 and 0.05

Another case is an experiment to separate sampling and testing errors in which several samples are taken and each sample is tested in duplicate.

Two common strategic errors when conducting interlaboratory trials are to conduct such a trial before studying the sensitivity of the test method to its parameters and to conduct such a trial before the participating laboratories are familiar with the test method.

Another type of design that is very useful is a staggered nested design as discussed in Section 6.3.10 on page 170. Such experiments do not have as much duplication at the lowest level.

Example: Checking bias and precision of iron ore sample preparation and testing

In the sampling of large quantities of iron ore there are usually many sampling operations. The first step is the taking of some increments from a stream of ore as it is being moved onto a stockpile, into a ship, onto a train, or for some other purpose. The primary samples are often very large and consist of coarse material. There are often several stages of subsampling and crushing or grinding before approximately a gram of ore in the form of a powder with a topsize of 100 microns is used to make a bead which is chemically analysed using an X-ray fluorescence machine.

It is usual to distinguish between 'sampling' and 'sample preparation'. Those operations which happen at essentially the same time and place as the primary sampling operation are called sampling. Those operations which happen later at a sample preparation plant are called sample preparation.

The bias and precision of routine sample preparation, and the bias and precision of routine testing were estimated by comparing the routine procedures to careful sample preparation and testing by experts. For each of a moderate number (say, 20) of samples the following results were obtained.

DECIDE ON A STRATEGY

- The routine result, y_R^R, was obtained by sample preparation and testing in the usual way. Note that the bucketful of material which was taken as the start of sample preparation was reconstituted by collecting all material which did not become part of the final small jar of 100 micron powder.
- An expert did careful testing in duplicate by preparing two beads from the jar of 100 micron powder which had been routinely prepared. Let us refer to these results as y_1^R and y_2^R, the superscripts indicating that the sample preparation was routine and the subscripts indicating expert results 1 and 2.
- The expert did careful sample preparation in duplicate to prepare two jars of 100 micron powder, then did careful testing in duplicate on powder from each of these jars. Let us refer to these results as y_1^1, y_2^1, y_1^2, and y_2^2.

Now the differences $y_1^R - y_2^R$, $y_1^1 - y_2^1$ and $y_1^2 - y_2^2$ provided plenty of information about the precision of expert testing. There is information about the precision of expert sample preparation in the differences $\frac{1}{2}(y_1^1 + y_2^1) - \frac{1}{2}(y_1^2 + y_2^2)$. All of these differences should have a mean of zero since there is no reason for the expert's first duplicate to give a higher value than the second duplicate.

Now the average of the distribution of the values of

$$\tfrac{1}{2}(y_1^R + y_2^R) - \tfrac{1}{4}(y_1^1 + y_2^1 + y_1^2 + y_2^2)$$

is the bias of the routine sample preparation procedure. The variance of the values can be expressed as a linear combination of the variances associated with routine and expert sample preparation and testing. Similarly, the average of the distribution of the values of

$$y_R^R - \tfrac{1}{2}(y_1^R + y_2^R)$$

is the bias of the routine measurement procedure, and the variance of the values can be expressed as another linear combination of the variances associated with routine and expert sample preparation and testing. This provides enough information to estimate all of the quantities of interest.

4.4 MAIN EFFECTS EXPERIMENTS

This section deals with strategies for estimating the effects of large numbers of factors when most of those effects can be adequately approximated by reasonably simple models. Such experiments cannot usefully be discussed without considering background information.

'Main effects experiments' are trials run for the purpose of estimating the influences of controllable factors on measurable responses, when we are primarily interested in the crude, overall effects of those factors. We wish to contrast them with two other types of experiments.

- Response surface experiments in which the effects of controllable factors are to be investigated in greater detail. In particular, with response surface experiments it is common to investigate many interactions between variables and to look at non linearity in the effects of factors.
- Experiments conducted with the primary aim of investigating components of variance.

4.4.1 Uses for Main Effects Experiments

Main effects experiments are likely to be useful in several different types of situations.

We may suspect that some or most of the factors to be investigated are irrelevant. This is commonly referred to as 'factor screening'. The most interesting part of the analysis of such experiments is testing whether the estimated effects of factors are significantly different from zero.

We may be investigating a situation about which little is known and consider that even a crude investigation is likely to be worthwhile. Here, interest is concentrated on the estimates of the effects of the factors, with formal statistical analysis often being unimportant. One such situation is when we wish to set specifications for the inputs to some process. We need information about the sensitivity of the process outcomes to the process inputs in order to set rational specifications. Often, little is known about such process sensitivities and a small amount of experimentation to help set specifications is easily justified.

Perhaps we are investigating a situation about which a substantial amount is known, including the likelihood that a simple model using only main effects will provide a good fit to reality. In such circumstances we are interested in the effects of the factors, the precision of these estimates and the goodness of fit of the model.

We may be in the first stage of response surface methodology. In this case we need to check the goodness of the fit of the model in order to decide whether to get more data or to fit a more complicated model. We may need to estimate the local effects of the factors in order to choose the levels of the factors for our next experiment.

Interpreting the residual error from a main effects experiment

After a main effects experiment has been done, the accuracy with which predictions can be made is constrained by the possibility that there may be substantial interactions between some of the factors. When data from such experiments are analysed, interactions are sometimes regarded as part of the noise. It is very important to differentiate clearly in your mind between two measures of the noisiness or the reproducibility of the results.

The pure error is the amount of variation you would expect between repeats of the same experiment. The residual error from your model measures the average amount by which the predictions of the model deviate from experimental outcomes. When this measure is based on a fractional factorial experiment involving several factors, it can be expected to be larger than the pure error because it includes the interactions.

After a main effects experiment, it is often desirable to follow up with a confirmatory experiment. If the results from such an experiment agree with predictions, then this is generally adequate evidence that interactions are relatively negligible. If the confirmatory experiment does not agree with predictions then the conclusions are unreliable.

A good feature of main effects experiments is that the most promising control factors are likely to all have been included as formal factors in the experiment. When the experiment is run, there is less likelihood of tampering because the tampering handles have preassigned values.

DECIDE ON A STRATEGY

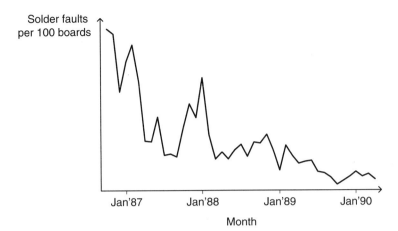

Figure 9. Plot of monthly data on the rate of solder faults

4.5 CASE STUDY: AN EXPERIMENT ON A WAVE SOLDERING MACHINE

This section discusses a substantial example of the programme of main effects experimentation. The outcome was not particularly successful.

4.5.1 Background

A company making spectrometers assembled and soldered its own circuit boards, using unloaded boards which it purchased. The unloaded boards were often referred to as printed circuit boards (PCBs), while the loaded boards were referred to as printed wiring boards (PWBs).

Solder faults had been a problem – see the graph of the monthly average number of solder faults per 100 boards in Figure 9. This data was readily available because there was 100% inspection of boards by an automatic testing machine. Note that the plant closed down for three weeks over Christmas and the soldering equipment had to be recommissioned.

4.5.2 Getting Started

Management was conscious of wanting to get the people who operate the wave soldering machine to be part of the improvement process, so those people were asked to draw up a list of factors

The initial list of things thought likely to affect wave soldering machine performance was as follows.

- Workers:
  - controller performance;
  - machine maintenance;
  - machine cleaning (solder pot cleaning, dross removal);
  - flux height setting;

— handling of PWBs and components.
- Machinery:
    - solder pre-heat temperature;
    - solder temperature;
    - conveyor speed;
    - time in solder wave;
    - wave dynamics (omega wave frequency);
    - wave height;
    - dross formation;
    - temperature error;
    - contamination from mask;
    - air contamination;
    - component movement;
    - oil in solder bath;
    - flux contamination.
- Materials:
    - flux activity (density);
    - contaminated components;
    - contaminated PWBs;
    - warped PWBs (finger tension);
    - size of PWB;
    - density of components on PWB;
    - metallurgy of PWB;
    - voids in plating;
    - solder alloy % tin;
    - solder contamination;
    - whether solder mask is cured.
- Management:
    - lead placement:
    - lead/hole ratio;
    - PWB orientation;
    - PWB art geometry;
    - lead length;
    - environment;
    - immersion depth of PWB into solder wave.

4.5.3 Selecting Factors and Levels

A small committee went through the list of potential factors, trying to classify the candidate factor into four categories.
- Factors that can be readily adjusted and which we would like to vary in the experiment.
- Factors which are important but which cannot be easily set to predetermined levels. Such factors are likely to be good explanatory variables, but be excluded from the experiment in favour of factors which can be readily adjusted.

Table 1. Experimental design and results for an experiment on a wave soldering machine

	2-level factors			3-level factors								Results		Comments	
Run	Flux contam.	Solder pot	Sold. oil clean used?	Pre-heat °C	Sold. temp. °C	Conv. speed m/s	Ω wave freq.	λ	Finger spacing	Flux dens.	Type of PWB	Air knife press.	Number of defects	Visual quality	Date	
1	new flux	before	yes	380	230	0.85	60	1400	spacer 1	900	1017	10	0	10	6/2/90	
2	new flux	before	yes	380	230	0.85	20	1500	spacer 1	920	1171	30	5 SD	10	2/2/90	
3	new flux	before	no	380	230	1.00	20	1300	no spacer	910	939	10	5 SD	6	9/2/90	1
4	new flux	after	yes	380	230	1.20	100	1300	spacer 2	900	1171	20	0	10	1/2/90	
5	new flux	after	no	380	240	1.20	100	1300	spacer 1	910	1017	30	0	10	26/1/90	
6	new flux	after	no	380	240	1.20	60	1500	spacer 2	920	939	30	2	8	1/2/90	2
7	old flux	before	no	380	240	0.85	20	1500	no spacer	900	939	20	0	10	8/12/89	
8	old flux	before	no	380	240	1.00	100	1400	no spacer	920	1171	10	0	10	8/12/89	3
8	old flux	before	no	380	240	1.00	100	1400	no spacer	920	1171	10	0	10	8/12/89	3
9	old flux	before	yes	380	250	1.00	100	1400	spacer 1	900	939	30	0	10	30/11/89	
10	old flux	after	no	380	250	1.00	20	1300	spacer 2	920	1017	20	0	10	22/11/89	
11	old flux	after	yes	380	250	1.20	60	1500	no spacer	910	1017	10	1 SD	9	17/11/89	4
12	old flux	after	yes	380	250	0.85	60	1400	spacer 2	910	1171	20	0	8	28/11/89	5
13	new flux	before	yes	420	240	1.00	100	1500	spacer 2	910	939	20	0	10	31/1/90	
14	new flux	before	yes	420	240	1.00	60	1300	spacer 2	900	1017	10	0	10	31/1/90	
15	new flux	before	no	420	240	1.20	20	1400	spacer 1	920	1171	20	1 SD	10	15/2/90	
16	new flux	after	yes	420	240	0.85	20	1400	no spacer	910	1017	30	0	10	13/2/90	
17	new flux	after	no	420	250	0.85	20	1400	spacer 2	920	939	10	1	10	30/1/90	
17	new flux	after	no	420	250	0.85	20	1400	spacer 2	920	939	10	1	10	30/1/90	
18	new flux	after	yes	420	250	0.85	100	1300	no spacer	900	1171	20	0	10	30/1/90	
19	old flux	before	no	420	250	1.00	60	1300	spacer 1	910	1171	30	1	9	28/11/89	6
20	old flux	before	no	420	250	1.20	20	1500	spacer 1	900	1017	20	0	10	27/11/89	
21	old flux	before	yes	420	230	1.20	20	1500	spacer 2	900	1171	10	0	9	18/11/89	7
22	old flux	after	no	420	230	1.20	60	1400	no spacer	900	939	30	0	5/9=8	27/11/89	8
23	old flux	after	yes	420	230	0.85	100	1300	spacer 1	920	939	20	0	10	18/11/89	
24	old flux	after	yes	420	230	1.00	100	1500	no spacer	920	1017	30	0	7	25/11/89	9
25	new flux	before	yes	460	250	1.20	20	1300	no spacer	920	1171	30	0	7	6/2/90	10
26	new flux	before	yes	460	250	1.20	100	1400	no spacer	910	939	20	0	10	6/2/90	
27	new flux	before	no	460	250	0.85	100	1500	spacer 2	900	1017	30	0	10	15/2/90	
28	new flux	after	yes	460	250	1.00	60	1500	spacer 1	920	939	10	0	10	1/2/90	
29	new flux	after	no	460	230	1.00	60	1500	no spacer	900	1171	20	0	10	26/1/90	
30	new flux	after	no	460	230	1.00	20	1400	spacer 1	910	1017	20	0	10	7/2/90	
31	old flux	before	no	460	230	1.20	100	1400	spacer 2	920	1017	10	0	10	8/12/89	11
32	old flux	before	no	460	230	0.85	60	1300	spacer 2	910	939	30	0	8	27/11/89	12
33	old flux	before	yes	460	240	0.85	60	1300	no spacer	920	1017	20	0	10	5/12/89	13
34	old flux	after	no	460	240	0.85	100	1500	spacer 1	910	1171	10	1	9	6/12/89	
35	old flux	after	yes	460	240	1.00	60	1400	spacer 2	900	1171	30	5	5/9=8	15/12/89	14
36	old flux	after	yes	460	240	1.20	20	1300	spacer 1	900	939	10	0	10	30/11/89	

Notes: In the quantity of defects column, SD is an abbreviation for 'solder dags'. In the visual quality column, the entry '5/9=8' appears where an initial rating of 5 was rated again as 9 and an adjudication process led to the rating of 8 being used for data analysis. The comments given in the last column of the table were as follows.

1. Some holes/legs not soldered at side
2. Some solder spots on resistors
3. Solder on top of first board
4. Small solder splashes on resistors
5. Some solder spots on resistors
6. One blow hole
7. Empty holes at edges
8. Many blowholes for one board only. Other 2 very good.
9. Solder splashes on resistors
10. Some holes at edges not soldered
11. Boards look good
12. Some dross on board
13. First run on 17/11/89 gave 'quantity of defects: 17 solder dags; visual quality: 5'. It was repeated because there had been water in an air line.
14. Many blow holes on one board. Other 2 very good.

- Factors which will be standardized in order to reduce variation from that source.
- Things which will be neglected, at least for this first experiment.

We tried to be explicit about what was meant by factors and what operational definitions might be used to define their levels. Some extra factors were added. Thought was given about the number of levels and the choice of levels. The three types of circuit boards included in the experiment were common types of typical sizes.

Adjusting λ on the wave soldering machine adjusts the speed of a solder pump and thereby affects the size of the solder wave. Normal operating procedure includes frequent adjustment of λ between types of boards. The λ values were selected hoping for simplification of operating instructions, with λ not varying.

4.5.4 What to Measure

In addition to recording the number of solder faults found by the automatic testing machine, a visual assessment of board quality was made.

4.5.5 Design

The design was an orthogonal array with 36 runs as given in the table. It was chosen largely because a similar design had been used in Lin and Kackar (1986). The prior usage made the experimental design easy for the people running the process to accept. Thirteen errors in the orthogonal array as printed in that paper were corrected before using it.

The order of runs was not explicitly randomized. However, the assignment of actual factor levels to formal levels was randomized (or, at least, haphazard). All runs with contaminated flux were done first.

4.5.6 Running the Experiment

Experimental runs were done when it was convenient to fit them in with normal production. Boards soldered under experimental conditions were visually inspected, but otherwise were treated just the same way as all other boards. They were soldered in batches of three. The number of defects was recorded as the total for the three boards and the visual quality recorded was based on assessment of all three boards.

4.5.7 Analysis

Results are given in Table 1, together with the dates of experimental runs. Four extra dependent variables were constructed from the comments. These variables were assigned values 0 or 1 depending on whether or not there were any solder dags, joints not soldered, spots or splashes, or blow holes.

Figure 10 was the primary form of presentation of the analysis. Settings were decided on the basis of the graphs. The variable 'visual quality' has been replaced by its negative, 'visual badness' so that lower is better on all graphs. This makes the task of choosing levels for the factors from the graphs less confusing.

DECIDE ON A STRATEGY

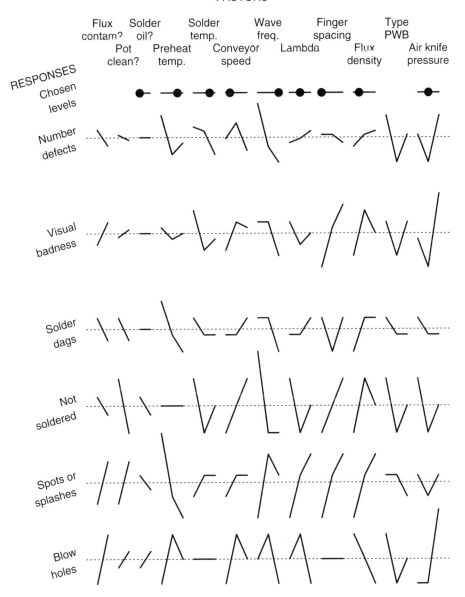

Figure 10. Estimated effects of various factors for an experiment on a wave soldering machine. This figure was used to choose levels for the factors. The process was run at the chosen levels from the start of March 1990

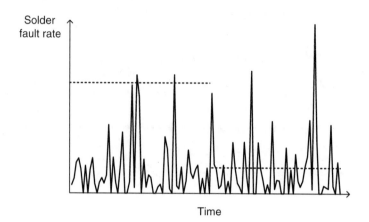

Figure 11. Daily data on solder fault rate from 27 November 1989 to 21 May 1990. The dotted lines show target values which were referred to as the 'control limit'

Spacings on the vertical scale are such that the standard errors for estimated effects correspond to the same distance for each of the six dependent variables. This is often a reasonable way to scale different variables, but is somewhat unreliable here because highly saturated models are being fitted so the standard errors are based on a few degrees of freedom.

4.5.8 Impact of this Work

This experiment didn't end up providing much useful information. The main useful outcome was probably a slight change in attitudes.

Data on daily solder faults for the period 27 November 1989 to 21 May 1990 is given in Figure 11. The daily data on solder faults is treated to some extent as a control chart. The dashed line shows a 'control limit' which was changed arbitrarily at the end of February 1990. Much of the variation is due to changes between the types of boards being soldered. It is thought that some variation is due to variation in the unloaded boards. Perhaps a future experiment might involve the board manufacturer.

Preheat temperature was subsequently lowered by 5 °C in order to avoid a problem. For some types of boards not included in the experiment a solder mask is used. This melts if too high a preheat temperature is used. The company now regularly calibrate the temperature gauges on their wave soldering machine.

A major conclusion from the experiment was that the settings of the machine are not as critical as they had been thought to be.

4.5.9 Lessons

Think more about what to measure. Here we would expect simpler (more nearly additive) effects for the measures which were constructed from the comments. Measures appropriate for routine monitoring are often not appropriate for experimentation. More costly or difficult measures are often justified for the duration

DECIDE ON A STRATEGY

of an experiment. The use of test boards should have been more seriously considered. The company has ordered art work for manufacture of test boards.

The involvement of appropriate people is important. The process of planning an experiment may be as important as the experiment itself. Here the employees of the company felt that they had learnt quite a lot merely by being involved in planning the experiment. They have discovered that the sides of their wave soldering machine are of slightly different heights. They are trying out stainless steel clips to hold large boards rigid–preventing the bowing which otherwise occurs when the top and bottom faces of boards are unequally heated. They have found that the temperature gauges need to be checked periodically.

The possible interaction between the factors which affect time in the solder wave ought to have been considered in the choice of experimental design.

4.6 RESPONSE SURFACE METHODOLOGY

This section discusses strategies for use when factors are generally already identified but a fairly precise idea of the trend is required. Response surface methodology can be discussed in the abstract, without considering the background knowledge available to the experimenter.

4.6.1 Types of Response Surface Methodology

There are two major issues which influence the overall strategy adopted in response surface problems.

Let us use the term 'on-line' experimentation to mean trials using full-scale production plant while endeavouring to continue to run the plant at normal production rates and to sell the output from the plant for the usual price. Let us use the term 'off-line' experimentation to mean experiments conducted without production pressures. This includes laboratory-scale experiments and commissioning trials.

The first issue is that on-line experimentation requires a more conservative approach than does off-line experimentation. This is because the cost structures are different. For on-line experiments, the largest potential costs are costs associated with producing substandard output or running at less than normal production rates. Extra replicates are not expensive because the production plant must be run in order to produce products for sale. For off-line experiments, the costs are those associated with setting-up, raw materials, measurement and the lost opportunities of not doing other experiments. Extra replicates are expensive.

The second issue is whether we wish simply to find a set of optimum conditions or whether we need to know the shape of the response surface over a substantial region. If there is an obvious function to be optimized it is often enough to find a set of optimum conditions. This includes the situation where a composite response can be calculated, possibly as a weighted sum of the squared deviations from target for each of several simple responses.

However, if there are several important characteristics of the output of a process which might vary in importance then it may be necessary to explore the response surfaces for the various responses of interest over a wide range.

4.6.2 Structure of the Remainder of the Notes on Response Surface Methodology

- Standard methodology when have a single response to be optimized off-line.
- Standard methodology when have a single response to be optimized on-line. This is called EVOP.
- Methodology when have several variables. This involves modelling each response separately, and using the models to make decisions. If there are only two independent variables, the contour plots of the fitted surfaces for various responses can be displayed on an overhead projector.

4.6.3 Types of Experimental Designs Used

The designs used are 2^2 or 2^3 factorial designs or a half fraction of a 2^3. Frequently centre points and star points are added in order to aid the fitting of response surfaces.

- 'Centre points' are points such that the levels of all factors are at the mid-point of levels used in the factorial design.
- 'Star points' are points such that all factors except one are at the mid-point of levels used in the factorial design. That one factor is generally set to outside the range used in the factorial part of the experiment.
- The number of centre points should be at least one. If the design is blocked then it is common practice to include a centre point in each block. Replication of the centre point gives information about repeatability where it is likely to be most relevant. There is little point in having more than four centre points.
- The position of the star points is often chosen so that the design is rotatable in the sense that it is equally informative about the response surface in all directions from the centre of the design.

Both of these last two choices are not very important in my opinion.

4.6.4 Fitting of Response Surfaces

This is usually done by regression analysis. Display is often conveniently done by plotting contour maps – character plots are generally adequate.

Fitting of response surfaces is often done sequentially. If there were three or more independent variables the following steps might be considered.

1. Half of 2^3 plus 2 centre points. You may stop here if in the preliminary stage of investigation where first-order models and screening designs give big benefits.
2. Move design.
3. Rescale design.
4. Drop or add factors.
5. Add other half of 2^3 plus 2 centre points.
6. Augment by adding star points.
7. Replicate (e.g. $\frac{1}{2}$ factorial to $1\frac{1}{2}$ factorial).

You may need more detailed experimentation near the maximum. A common mistake to be wary of is that people often choose to experiment over too small a region. In the neighbourhood of a maximum it is necessary to get good estimates of the second

DECIDE ON A STRATEGY

order behaviour of the response surface, so points should be chosen with factor levels sufficiently spread out that the likely fractional error in estimates of second-order terms is small.

Box, Hunter and Hunter (1978, pages 536–7) gave a flow diagram illustrating how response surface methods typically ought to proceed. A similar description is given below.

First-order model fitting phase

- Design and conduct the first-order experiment involving all the independent variables. The design chosen should be capable of being augmented to give a good second-order design. Analyse the results.
- If fit is inadequate, consider transformations of independent variables or responses to improve fit. Then, if fit is still inadequate, start on second-order phase.
- If precision is inadequate then start this phase again with different ranges for independent variables, additional replication, or making use of blocking variables.
- Determine direction of steepest ascent for chosen objective. Note that there is some arbitrariness about the definition of the direction of steepest ascent when the adjustable factors are in different units. The most satisfactory way of resolving this is to express factors in standardized units which each correspond to a moderate amount of adjustment. For instance, you might believe that 15 g of ingredient X is about as important as 10°C which is about as important as 2 minutes reaction time.
- Design and conduct one or more one-dimensional experiments. Analyse the results to determine the optimum within a one-dimensional search region.
- Reconsider the objective of your investigation – change it if necessary.
- If it is judged likely to be worthwhile then start this phase again using the newly obtained starting point. Otherwise, write report using data so far obtained.

Second-order model fitting phase

- Design and conduct second-order experiment involving all the independent variables. Make use of experimental results already obtained where they are near enough to be relevant. Analyse the results fitting a quadratic surface.
- If fit is inadequate, consider transformations of independent variables or responses to improve fit.
- If precision is inadequate then start this phase again with different ranges for independent variables, additional replication, or making use of blocking variables.
- Find the optimum of estimated response surface. Possibly draw contour plots of objective function and/or perform canonical analysis. Possibly also re-fit the response surface having defined its parameters relative to the location of the estimated optimum and the orientation of the eigenvectors of the second derivatives at the estimated optimum. This is called the 'double linear regression method' by Bisgaard and Ankenman (1996). It allows convenient calculation of confidence intervals of features of the response surface which are likely to be of interest.
- If it's felt necessary, perform additional runs in order to reduce uncertainty about the location of the optimum or other feature of response surface.

- Reconsider the objective of investigation – if necessary, redefine it and recommence at an appropriate point.

4.6.5 Strategies for Use with Many Responses or Many Objectives

The general methods are to model each response separately, and to construct a composite response.

Constructing a composite response is a practical approach when the relative importance of the various responses is likely to be reasonably constant over the foreseeable future. Such a composite response often amounts to expressing everything in current dollar terms.

The basic approach is to quantify the unsatisfactoriness of having other than the ideal value for each of the responses. The total of these measures of unsatisfactoriness is a measure of how unsatisfactory a set of operating conditions seem to be. The composite response may be taken to be the total of the measures, in which case a minimum should be sought for this one-dimensional composite response. Alternatively, the composite response could be taken to be some constant minus the total, so that a maximum is to be sought.

The advantage of constructing a composite response is that the response surface only needs to be studied near to its optimum. The methods discussed in an earlier section of these notes may be used.

Modelling each response separately is the subject of this section. It has some advantages over constructing a composite response. The results remain applicable even if the relative importance of the various responses changes, and even if the target values for some individual responses are changed. Simple response surfaces are likely to provide an adequate fit to individual responses, but more complicated surfaces are likely to be needed for a composite response. Background knowledge and intuition are more useful for checking and interpreting results for one response at a time.

Its disadvantages are that there are more response surfaces to report and that they must generally be investigated over a larger region.

For two independent variables, the contour plots of the fitted surfaces for various responses can be displayed on an overhead projector. If there are more than two independent variables then other methods of summarizing the results must be found.

4.6.6 Warnings and Pointers

It is commonly reasonable to assume that experiments which you are planning are not the first experiments to have been done on the process which you are investigating. One-factor-at-a-time experiments would have found a point on a ridge. Therefore it is likely that the current conditions are on a ridge. The consequence of this is that you might bypass the first-order model fitting phase, or at least be very concerned about the second-order experiment which might be produced by augmenting the first-order experiment.

Suppose you are concerned about slowly-varying sources of error. Perhaps the quality of a raw material may be changing or a piece of test equipment does not hold its calibration but tends to drift. In such circumstances it may be useful to block your experiment. This assumes that trials within a block are consistent but that there

DECIDE ON A STRATEGY

may be step changes from block to block. This is not a perfect model but is good enough for practical purposes. It is robust.

Sometimes the region over which the true response can be well approximated by a quadratic surface is small relative to the amounts by which the levels of the factors need to be spread in order to fit the quadratic surface with few observations. There are two options.

- You can use a large number of observations taken within a region which seems likely to include the location of the optimum set of conditions and over which a quadratic surface appears to be a good approximation to the true response surface. This has the disadvantage of being likely to require more observations than the alternative.
- You can fit a mathematical form to the response surface which is more complicated than a quadratic surface, using observations over a larger region. The effectiveness of this approach depends on your choice of mathematical form and of region.

Response surface methodology can be usefully practised using simulated data. One person can choose a mathematical function which could conceivably describe the trend and choose a model for generating noise to add to the trend to provide the simulated data. Another person, the one practising response surface methodology, must specify the conditions for which they would like some data, must analyse the data, and must decide whether to get more data or to stop because further precision is finding the best conditions cannot be justified. When I asked a group of my colleagues, all highly trained in mathematics and statistics, to play such a game, I generated highly correlated noise. They performed poorly, so I recommend using white noise, at least for the first time that people play this game.

4.6.7 EVolutionary OPeration

EVOP is when a process is run in such a way that it generates information on the product as well as information on how to improve the manufacturing process. The types of experimental designs used are the same as discussed in the section of response surfaces, with the simple designs being most common.

EVOP differs from standard response surface methodology in that the experimentation is more conservative. The factors are never changed in large steps. This is appropriate in two situations.

- EVOP is appropriate when the main cost of observations is the possible loss of production. The cost of changing the levels of the factors being investigated is assumed to be small and the cost of measuring the characteristics of the output is also assumed to be small – because they would have been measured anyway.
- EVOP is also appropriate when the shape of the true response is not well approximated by a quadratic over a sufficiently large domain for estimation of the response surface in the region of the optimum.

EVOP provides a framework for never-ending improvement of a process. The steps involved are as follows (essentially following Hunter (1988)).

1. Obtain agreement and support from production management.
2. Survey company reports and open literature on the process. Study cost, yield and production records.

3. Learn about EVOP. Hold training sessions for people involved.
4. Select two or three factors which seem likely to influence the most important response.
5. Using a simple experimental design such as a 2^2 or 2^3 plus centre point, vary these factors about currently accepted conditions in small steps.
6. Replicate this simple design, estimating the effects after completing each replicate beyond the first. If one or more effects are statistically significant, move in the direction indicated and start a new phase of EVOP, possibly with new factors or new ranges for factors.
7. If no factor is statistically significant after, say, 8 replications then change the ranges or select new variables.
8. When near enough to an optimum, run a new phase of EVOP with a different set of factors.

See Box and Draper (1969) for a detailed description of EVOP and Hunter and Kittrell (1966) for a review of some applications. In my opinion, Hunter and Kittrell's review provides surprisingly little support for the idea that EVOP is a very effective technique. I would have expected such a paper to concentrate on cases for which EVOP was an astounding success, so it appears that there are few such cases. I suspect that EVOP is only effective at finding ridges, when one-factor-at-a-time experiments would also have been effective.

The most important principle to remember is that the way to learn is by trying out alternatives.

One criticism is that EVOP is not particularly efficient for moving along ridges. It is better if one can allow a reorientation of design. The important part of EVOP is to note that information can be obtained without unacceptable risk of producing an off-spec product.

4.6.8 When to use EVOP

The advantages of EVOP over response surface designs which make larger modifications to the controllable variables are as follows.

- There is less risk of producing a product which is out of specification. If this advantage is not very important then EVOP should not be considered.
- It is natural to block the runs.
- Progress is not much affected by the shape of the true response surface far from the optimum. In contrast, response surface designs depend on the assumption that a quadratic response surface is a reasonable approximation over the (generally larger) experimental region used.
- The process workers tend to feel that they own any improvement achieved, and therefore try to ensure that minor difficulties are overcome.

Its disadvantages are these.

- When the cost of experimenting is primarily the cost of lost production, EVOP is a more expensive way of finding out about second derivatives than using response surface designs.

DECIDE ON A STRATEGY

- EVOP requires more supervisory time, checking the changes to process parameters are made as designed.
- EVOP requires more analysis effort.

4.6.9 Example with Two Factors

This example is included here in order to illustrate reponse surface methodology with more than one response variable. It has been taken from Snee, Hare and Trout (1985). If referring to that book, note that the axes on some contour plots appear to have been incorrectly transposed.

Gentamicin is a therapeutic antibiotic drug of last resort. When it is administered, its concentration in the blood should be kept below $12\,\mu\mathrm{g/ml}$ in order to avoid kidney damage and kept above $2\,\mu\mathrm{g/ml}$ in order to be effective.

A method of determining the concentration of gentamicin in the blood is to add some reagents to a sample of blood and measure the absorbance of selected wavelengths of light. The absorbance reduces more quickly if there is more gentamicin because the reaction involving gentamicin removes a chemical which absorbs light.

Two factors describing the reagent mixture can be varied:

- the particle reagent volume (ml) to be referred to as x and used as the horizontal axis on Figures 12 to 16, and
- the antibody volume (ml) to be referred to as y and used as the vertical axis on Figures 12 to 16.

The measurements made for each reagent mixture were the initial rate of change of absorbance for calibrated samples with $16\,\mu\mathrm{g/ml}$ and $4\,\mu\mathrm{g/ml}$ of gentamicin and for a sample with no gentamicin (a blank). Let us denote these by R_{16}, R_4 and R_0. In addition, the final absorbance for the blank was recorded. Let us denote this by F.

The responses of interest are:

- the blank absorbance rate, R_0, which needs to be greater than 300;
- the final blank absorbance, F, which should be less than 0.8;
- the difference between the absorbance rates for the blank and the sample with $16\,\mu\mathrm{g/ml}$ of gentamicin, $R_{16} - R_4$, (referred to as the separation, s) which needs to be greater than 200, and;
- a ratio, $r = (R_0 - R_4)/(R_0 - R_{16})$ which needs to be between 0.4 and 0.5.

Three replicates were conducted, and the medians of the results obtained were as given in Table 2. The calculated responses s and r are also given.

Response surfaces were fitted to each of the four responses of interest, using $X = (x-35)/15$ and $Y = (y-25)/15$. Note that both X and Y are therefore between -1 and $+1$. The models fitted were products of quadratics in X and quadratics in Y. There is enough flexibility in this class of models to provide a perfect fit to the sets of nine medians. The possibility that this might be overfitting was considered by looking at the statistical significance of the the various terms using all replicates but was not discussed to my satisfaction. The coefficients of the fitted models are given in Table 3.

Contour plots of the fitted surfaces are given in Figures 12, 13, 14 and 15. These can be easily overlaid as illustrated in Figure 16 in order to find the satisfactory region.

Table 2. Response measures for gentamicin experiment

x	y	F	R_0	R_4	R_{16}	s	r
20	10	0.243	70	8	2	68	0.9118
35	10	0.379	310	241.5	29	281	0.2438
50	10	0.439	309	326	141	158	−0.1076
20	25	0.383	85	14	5	80	0.8875
35	25	0.620	485	319	53	432	0.3843
50	25	0.735	435	428	192	143	0.0490
20	40	0.511	83	18.5	9	74	0.8716
35	40	0.814	541	306	72	469	0.5011
50	40	0.954	432	529	240	192	−0.5052

Table 3. Estimated coefficients of models fitted to responses for gentamicin experiment

	F	R_0	s	r
(Const)	0.6200	485.0	432.0	0.3843
X	0.1760	175.0	31.5	−0.4193
Y	0.2175	115.5	94.0	0.1287
XY	0.0618	27.5	7.0	−0.0894
X^2	−0.0610	−225.0	−320.5	0.0840
Y^2	−0.0235	−59.5	−57.0	−0.0119
X^2Y	−0.0217	−81.5	−84.0	−0.2381
XY^2	−0.0163	−28.0	20.5	−0.1798
X^2Y^2	0.0012	23.0	68.5	−0.1638

- Shade the areas which are unacceptable for a particular response on a transparency of the contours of that response surface. This can be done in black and white by using different modes of shading for the different response surfaces, as has been done in the figures being discussed. The display will be easier to understand if different colours are used for the different responses.
- Overlay the transparencies, looking first for any region which is clear. This gives conditions which make all responses acceptable.
- Look for areas which are only unacceptable on one response. Consider whether the limits on that response are accurately determined.

4.7 CASE STUDY: A SIMULATION GAME

This section presents the results of following a number of different strategies for playing a simulated response surface game. This allows comparison of outcomes of using the various strategies. This would not be practical for a real situation, because it would be impossible to justify the expense of performing several sets of experiments after a process had already been investigated and a satisfactory approximation to the best operating conditions had been found.

DECIDE ON A STRATEGY

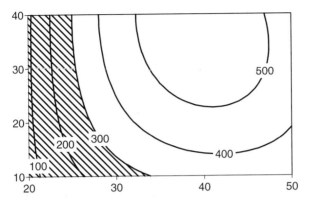

Figure 12. Contours of fitted response surface for blank absorbance rate, R_0. The shaded region gives unacceptable values for this response

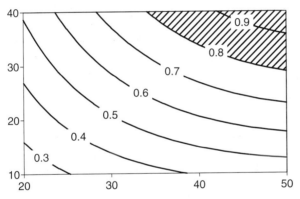

Figure 13. Contours of fitted response surface for final blank absorbance, F

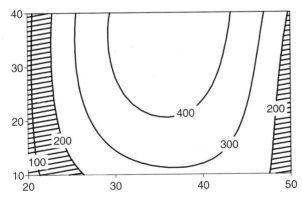

Figure 14. Contours of fitted response surface for separation, s

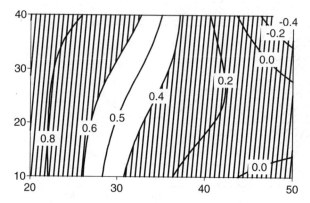

Figure 15. Contours of fitted response surface for ratio, r

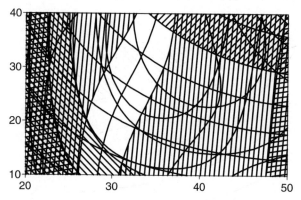

Figure 16. All four contour plots superimposed. The shading can be used to tell which responses have unacceptable values

4.7.1 The Simulated Situation

The mathematical formula used for the average response is

$$101 \exp\left[-0.0003 \left(x + 12\sin(\tfrac{x+y}{30}) - 60\right)^2 - 0.002 \left(0.8x + y + 9\sin(\tfrac{x-y}{20}) - 100\right)^2\right].$$

The noise added was uncorrelated, normally distributed with standard deviation 10.

Note that because the noise is uncorrelated we do not need to worry about randomization and we do not need to worry about combining one set of runs with another set done at a different time.

Figure 17 shows contours of the response surface. These contours are at heights of 10, 20, ..., 100. Some of them are labelled. The bold continuous line shows the mathematically natural ridge defined as being where the larger squared term is zero, namely

$$0.8x + y + 9\sin(\tfrac{x-y}{20}) - 100 = 0.$$

DECIDE ON A STRATEGY

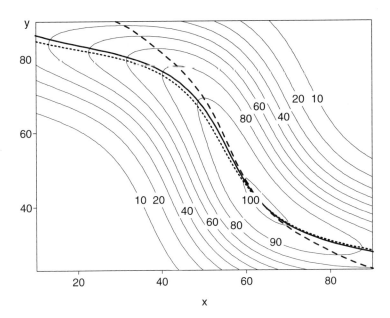

Figure 17. Contours of the true response surface used for a simulation game. Three lines are drawn to indicate the ridge of the response surface. The continuous line is the mathematically natural ridge. The dotted line shows the maximum y given x. The dashed line shows the maximum x given y

The dotted line shows the y values which maximize the mean response for fixed x. Tangents to the contours of the response surface are parallel to the y-axis at points on the dotted line. The dashed line shows the x values which maximize the mean response for fixed y. Tangents to the contours of the response surface are parallel to the x-axis at points on the dashed line.

Efforts to improve the average response will be started at the point (40,65), so far as possible.

4.7.2 Strict One-Factor-at-a-Time Approach

This approach was started by fixing x at 40 and varying y around the starting value of 65. Three runs were done at y values of 45, 55, 65, 75 and 85. Figure 18 shows the data obtained. There are two points with responses of 60.90 and 60.73 for $y = 85$ which appear as a single blob on this plot. The dashed line shows the true average response.

For each one-factor experiment, there are some decisions which must be made, with little or no information available with which to make them.

1. We have to decide on the spacing of the levels of the factor. For the first experiment, the spacing of 10 was decided arbitrarily. For subsequent one-factor experiments, it is possible to make use of the experience and information from earlier experiments to ensure that the spacing used is sensible.

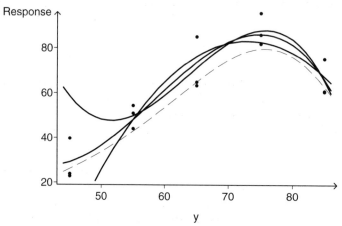

Figure 18. First one-factor-at-a-time experiment. The dots show the data obtained in this experiment. The dashed line shows the true average response. The continuous lines show three different curves fitted to various subsets of the points

2. We have to decide on the number of replications. For the first experiment, three replicates were used in order to obtain a moderate amount of information about the variability of the responses. For subsequent one-factor experiments, the number of replicates is taken to be larger when greater precision is required.
3. We have to decide what model to fit, and whether to use all of the data when fitting a model. For the first experiment, Figure 18 shows three fitted curves. One is a cubic fitted to all of the data. This curve is the middle of the three for most y. Its maximum is at $y = 74.8$. Another is a quadratic fitted to the data for $y > 50$. This curve is the lowest at $y = 50$. Its maximum is at $y = 72.9$. Another is a cubic fitted to the data for $y > 50$. This is the curve which is highest at $y = 45$. It has its maximum at $y = 75.7$.

   For most of the one-factor experiments, I used a cubic curve fitted using data from relatively close to the apparent optimum as judged from a graph. The reason for preferring cubics to quadratic curves is that they fit better when the decline from the maximum is more rapid on one side than on the other.

The second experiment was with $y = 75.7$, this being the optimum found on the previous step. x values of 20, 30, 40, 50 and 60 were tried. Three replicates were again used. The best x was found to be $x = 41.7$.

It might have been reasonable to decide to stop the sequence of one-factor-at-a-time experiments after this second experiment. After all, the starting value of $x = 40$ seems to be near enough to being optimal. However, I wanted to persist. I wanted to see whether a sequence of one-factor-at-a-time experiments can optimize my simulated process.

For any response surface that has a ridge, a first one-factor-at-a-time experiment starting from somewhere not on the ridge is very likely to find a point which is near to the ridge. A second one-factor-at-a-time experiment might suggest that another point

DECIDE ON A STRATEGY

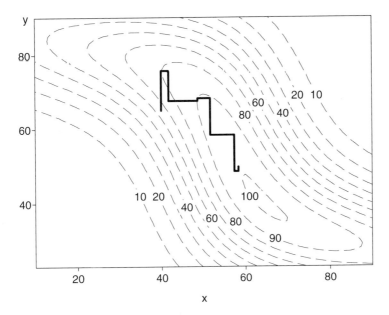

Figure 19. An outcome of playing a simulation game with a strict one-variable-at-a-time strategy. The contours of the true response surface are shown as dashed lines. The zig-zag solid line shows the progress of a sequence of one-factor-at-a-time experiments

near to the ridge is a little better, but is unlikely to make as dramatic an improvement as the first one-factor experiment.

The sequence of one-factor-at-a-time experiments is illustrated in Figure 19 by a solid zig-zag line. The x values which were held fixed in experiments 1, 3, ..., 13 were 40, 41.7, 41.8, 48.7, 51.7, 57.4, and 58.4. The y values which were held fixed in experiments 2, 4, ..., 12 were 75.7, 75.4, 67.6, 68.3, 58.4 and 48.6. The estimated best value for y according to the 13th experiment in which x was fixed at 58.4 was $y = 49.8$.

Looking at the path of successive estimated optimum points, it seems that most rapid progress towards the true optimum was made when an estimated optimum was far from the true ridge. The following one-factor experiment then found an estimated optimum which was near to the true ridge but substantially closer to the true global maximum.

A major difficulty with the strict one-factor-at-a-time approach is that it is very difficult to decide when to stop experimenting. I knew from previous experience that the method is good at finding ridges, but not very efficient at moving along them. So I kept experimenting, even though getting results like the following.

- Experiment 2 found that the best x was 41.7, which was very near to the starting value of 40.
- Experiment 3 found that the best y was 75.4, which was very near to the previous best value of 75.7.
- Experiment 4 found that the best x was 41.8, which was very near to the previous best value of 41.7.

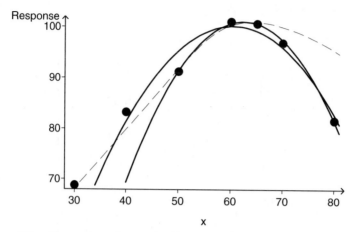

Figure 20. Illustration of an embedded one-factor-at-a-time experiment as an experiment over x which uses the estimated maximum over y as response variable. The dashed line shows the true average response, assuming that the best y is used for each x. The continuous lines show two different curves fitted to various subsets of the points

4.7.3 An Embedded One-Factor-at-a-Time Approach

Several one-factor-at-a-time experiments were done in which x was kept constant and y was varied. For each of these experiments we can estimate the maximum of the response and the y for which this maximum occurs. The individual experiments were each like the one illustrated in Figure 18.

The 'embedded one-factor-at-a-time approach' may in this case be regarded as a one-factor experiment over x which uses the estimated maximum over y as a response variable. It is a form of two-factor experiment.

The collection of one-factor-at-a-time experiments is illustrated in Figures 20 and 21. Figure 20 shows the experiment over x which uses the estimated maximum over y as response variable. Once a preliminary version of such a graph is obtained, it is then desirable to obtain greater precision for estimated maxima over y near the estimated optimum. This can be done by adding extra data points.

The number of trials used in this approach was 15, 15, 15, 25, 30, 25 and 15 for x being 30, 40, 50, 60, 65, 70 and 80, respectively. This is a total of 140 runs.

This is a quite good procedure, and could be used when there are several factors, provided that background information tells us which factor to maximize over.

4.7.4 An EVOP Approach

The crucial characteristic of EVOP is that small ranges are used for factors during any single experiment. This is appropriate for trials on production equipment.

Starting at (40,65) for (x, y), as for the other methods, the first round of EVOP experimentation used a range of ± 1 in each of the variables. Eight replicates of the set of five trials were run before a statistically significant effect was observed. Since the effect of x was statistically significant with response increasing with increasing

DECIDE ON A STRATEGY

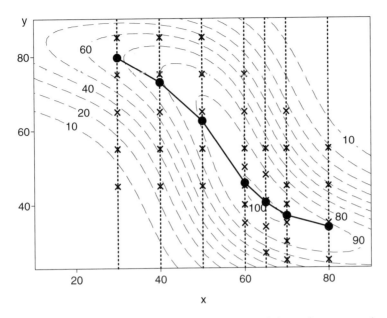

Figure 21. Illustration of some aspects of an embedded one-factor-at-a-time experiment. For each of the x values shown by a vertical dotted line, data was obtained at the points shown by crosses. Data was also obtained for $x = 80$, $y = 15$ which is outside the range of the plot. The location of the estimated maximum over y is shown by a solid circle. The line joining the solid circles shows the approximate location of the ridge of the simulated surface

x, the design was moved to the right. The distance moved was 2. This means that the left-hand side edge of the new experimental region corresponds to an edge of the previous experimental region.

Figure 22 shows all the stages of an attempt to maximize the response surface using an EVOP strategy. The first stage, described above, is shown as A, and subsequent stages are indicated as B, C, etc.

Ten replicates of the five trials in stage B were required before a statistically significant trend was found. Given that the direction of increasing mean response was consistent with the direction from the previous round of EVOP experimentation, it seemed appropriate to search cautiously in the direction (2,1). Since the mean response is larger than where we started, we can afford to be slightly less cautious, without much fear that the average will drop below that at our starting conditions.

Stage C did 13 replicates at the three points (43,66), (45,67) and (47,68), and led to the conclusion that it would be worthwhile to search further in the intended direction.

Stage D did 10 replicates at the three points (47,68), (49,69) and (51,70). A quadratic curve was fitted to the data of our one-dimensional search, indicating that the location of the best mean response is at about (48.2,68.6).

It was presumed to now be necessary to move along a ridge, with the ridge being approximately at right angles to the direction of steepest ascent which was followed in order to get to the ridge. A six-point design was used to attempt to find out whether

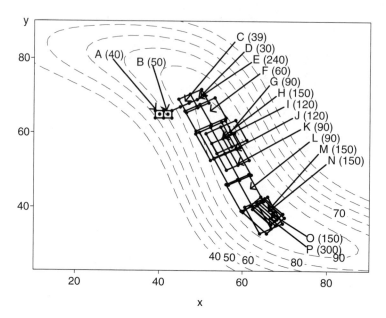

Figure 22. An outcome of playing a simulation game with an EVOP strategy. The dashed lines show contours of the true response surface. The points used in stages A to P are drawn with arrows pointing towards the middles of the sets of design points used. The numbers in parentheses indicate the numbers of runs used in the various stages

it was desirable to move in one direction or the other along the ridge, while ensuring that the search did not stray very far from the ridge.

For stage E of the EVOP approach, six points, as shown in Figure 22 were used. The points were at distances −2, 0 and 2 from the centre of the design, (48.2,68.6), in the direction (2,1) of the previous search and at distances −2 and +2 square to that direction.

Forty replicates of design E were run before it became apparent that the response surface increased significantly in one direction along the apparent ridge. The locations of the points of the design and the average responses are shown in Figure 23 The type of model being fitted had a quadratic trend in direction (2,1) and a linear trend square to that direction, which is direction (−1,2).

It appeared to be sensible to move the centre of the search region along the direction of the ridge, (1,−2), and a small amount sideways to the apparent ridge. It also seemed desirable to make the design a little larger so that fewer replications might be required before a statistically significant effect was likely to be found. These considerations led to stage F as shown on Figure 22.

Similar issues were considered after each subsequent stage of EVOP experimentation. Figure 22 indicates in parentheses the total number of trials conducted for each stage.

After stage M, it was considered appropriate to rotate the experimental region. The results for that stage were as shown on Figure 24, suggesting that the location of the

DECIDE ON A STRATEGY

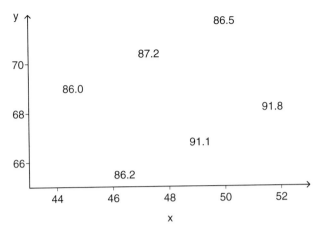

Figure 23. Average responses for stage E of EVOP strategy, printed at positions which indicate the values of x and y. The responses indicate that the average response increases in the direction of increasing x and decreasing y

ridge is near the centre of the upper set of three points but is not near the centre of the lower set of three points.

For stages N, O and P the rotated design was used. The spread of the design was decreased and the decision was made to stop experimenting because there had been little change in the estimated location of the maximum over four stages.

4.7.5 Using Sequential Response Surface Methodology

Here, I have endeavoured to follow textbook response surface methodology, endeavouring to minimize the number of experimental runs.

The first stage is to look for main effects. When the number of factors is small, it is common to use a design with a centre point so that it is possible to test for quadratic behaviour. The extra data point also provides an additional degree of freedom and thereby slightly improves the accuracy of early estimates or error variance.

Figure 25 shows the points used in the first stage as solid circles. They are more widely separated than the points used to start the EVOP strategy, so that it is reasonably likely that trends in the response surface can be reliably detected with a single replication. This is an appropriate strategy for experiments in a laboratory or in a pilot plant, but would not be appropriate when experimenting on a production facility and hoping to continue to be able to sell the output.

The data obtained were as follows:

x	y	Response
35	60	37.6
35	70	57.9
40	65	72.1
45	60	67.7
45	70	87.9

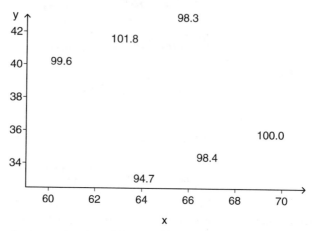

Figure 24. Average responses for stage M of EVOP strategy, printed at positions which indicate the values of x and y. The top of the ridge seems to be near the middle of the set of three numbers 99.6, 101.8 and 98.3, but to one side of the set of three numbers 94.7, 98.4 and 100.0

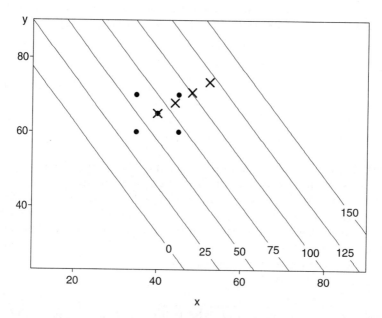

Figure 25. Illustration the first stage of playing a simulation game using response surface methodology. The points used in the first stage are shown as solid circles. Some contours of a linear response surface fitted to that data are shown as solid lines. The points to be used in the second stage are shown as crosses

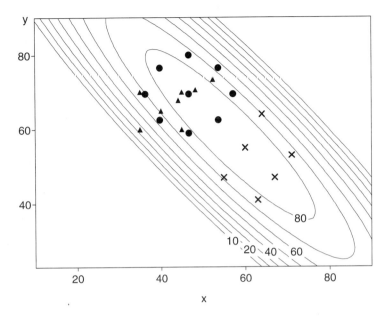

Figure 26. Illustration of the third stage of playing a simulation game using response surface methodology. The points used in the first and second stages are shown as solid triangles. The points used in the third stage are shown as solid circles. They constitute an octagon plus a centre point. The points proposed for the fourth stage are shown as crosses. The positions for these points were chosen considering the contours drawn which show the response surface fitted to the data from the third stage

A linear model was fitted to this data. Linear effects in x and y were both substantial. The effect in y was not statistically significant at the 95% confidence level, but this statistical test is not very important since there is no special credibility attached to the hypothesis that y has no effect on the response.

Figure 25 shows the contours of the linear response surface fitted using the data from the first stage. The second stage of standard response surface methodology is to look in the direction of steepest ascent. This could be done one point at a time or it could be done using an experiment involving several trials.

Four design points were used along a line in the direction of steepest ascent as suggested by the first phase. Three replicate trials were used for each of these design points. A quadratic curve was fitted to these twelve data points. It had its maximum at $x = 46.7$, $y = 69.5$. This is expected to be a point on a ridge of the response surface. It might happen to be near the global maximum, but this cannot be assumed.

The third stage of response surface methodology is to do enough trials so that it is possible to fit a quadratic surface in several variables (here only two) so that we can find the direction of a ridge. However, we should allow for the possibility that the optimum has already been (nearly) found.

The design used was a central composite design, which was eight points almost on a circle plus a centre point. These are shown in Figure 26 as solid circles.

It was found necessary to replicate the design for the third stage. The quadratic surface fitted to the data after two replications was saddle-shaped, suggesting that both directions along the ridge would give better responses than the region of the data. Four replicates were taken altogether, in order to get enough information about the shape of the response surface in the direction of the ridge.

A good procedure for checking the proposed spread of design points in the direction of steepest ascent would have been to simulate some possible responses using the quadratic surface fitted during the second stage as the average response and adding pseudorandom noise which has variance equal to the current best estimate of the noise variance. Such simulated data could be used to fit a quadratic surface, checking that the quadratic term in the direction of steepest ascent should be likely to be statistically significant. If this is not the case then a greater spread of design points should be considered. This procedure was not followed.

Also with hindsight, it might have been reasonable to have the design points in stage three more spread out square to the direction of steepest ascent than in the direction of steepest ascent, using the logic that at most spots on a well-defined ridge the second derivative along the ridge is much smaller (in absolute value) than the second derivative square to the ridge.

Data was obtained from the simulation game and a quadratic response surface was fitted. This means fitting linear terms, quadratic terms and cross-products of pairs of variables, and can be done easily using multiple regression. The contours of the fitted quadratic surface are shown in Figure 26.

It is now desirable to move along the ridge towards the maximum. Six points were chosen for stage four, considering the fitted quadratic surface shown in Figure 26 and the points for which data has already been obtained. It is expected that data from the additional six points will be able to be combined with data from stage three in order to estimate the response surface over a region which is thought likely to include the location of the maximum. Note that when experimenting on real processes, it is usually desirable to repeat some runs between different stages of response surface methodology experimentation where regions of interest overlap, to guard against drift in the process mean due to changes in raw materials, wear in equipment or other forms of drift.

Figure 27 shows the state of knowledge after the fourth stage of playing the simulation game using response surface methodology. Two replicates were taken at each of the six design points shown as solid circles on Figure 26.

For stage five, three replicates were taken of each of five new design points and four of the design points used first in stage four. Figure 28 shows the state of knowledge after the fifth stage of playing the simulation game using response surface methodology. Note that the estimated location of the maximum is quite near to the true location of the maximum. However, a quadratic surface is not a very good fit to the true surface over a range as wide as the spread of the points used for fitting the quadratic surface, so greater precision could not be expected without using a narrower spread of design points. Many replicates would then be required for accurate fitting of the model.

It is worth noting that the total number of data points used in this approach was

$$5 + 4 \times 3 + 9 \times 4 + 6 \times 2 + 9 \times 3 = 92.$$

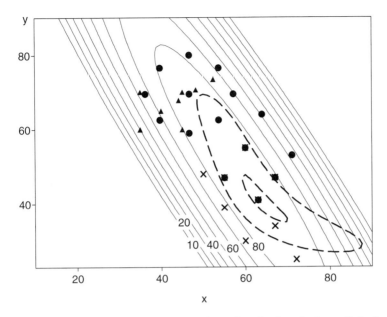

Figure 27. Contours of fitted response surface after the fourth stage of playing a simulation game using response surface methodology. The points used in the first and second stages are shown as solid triangles. The points used in the third and fourth stages are shown as solid circles. The five new points proposed for the fifth stage are shown as crosses. The positions for these points were chosen considering the contours drawn as continuous lines, which show the response surface fitted to the data from the third and fourth stages. The dashed lines are the level 90 and 100 contours for the true response surface

4.7.6 Using One-Hit Response Surface Methodology

An alternative to the approaches above is to have a single stage of experimentation. This is appropriate in practice only if you are confident that you already know the extent of the region of interest.

It is essential that enough data points be used to allow fitting of a surface of the complexity required. This can be checked approximately by looking at the number of parameters to be estimated in the model for the trend which is to be fitted. The amount of information available to describe the departures from the trend can be described by its number of degrees of freedom, which is the number of data points minus the number of parameters in the the model for the trend. It can be checked with greater certainty by generating some simulated data from your proposed experiment and using your intended data analysis procedure on the simulated data.

For the purpose of playing the simulation game, responses were collected for a 9×9 array of data points over the range 35 to 65 for x and over the range 40 to 70 for y. This range includes the standard starting conditions, $x = 40$ and $y = 65$, but is not centred about those standard starting conditions. It would not be appropriate to use a one-hit approach to experimentation unless you have a reasonable idea of where the interesting region lies, so use of the standard starting conditions was not sensible.

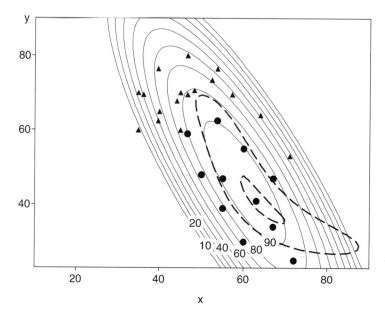

Figure 28. Contours of fitted response surface after the fifth stage of playing a simulation game using response surface methodology. The points used in the first and second stages are shown as solid triangles. The points used in the third, fourth and fifth stages are shown as solid circles if they were used for fitting the quadratic surface, otherwise they are shown as solid triangles. The continuous contours are for the fitted quadratic surface. The dashed contours are those for levels 90 and 100 of the true response surface

One reason for choosing this range is that, as can be seen from Figure 17, the true maximum is inside the region being investigated but is fairly close to the edge of the region being investigated.

Once the experiments have been done for a one-hit approach, the only remaining question is how to analyse the data. There are two parts to this question:

- Should all data be used for fitting a response surface, or should some data be downweighted or omitted?
- How complex a function should be fitted?

Figure 29 illustrates what can go wrong when a model with few parameters is fitted over a wide range. It shows as continuous lines the contours of a quadratic surface fitted to all 81 data points. The level 90 and 100 contours of the true response surface are shown as dashed lines for comparison.

The fitted quadratic surface is not a good approximation to the true response surface in the neighbourhood of the maximum. Such a quadratic surface (in two dimensions) can be described using only six parameters: a constant, two linear terms, two quadratic terms and one cross product term. This is not sufficient to describe the shape of the true surface over such a large region.

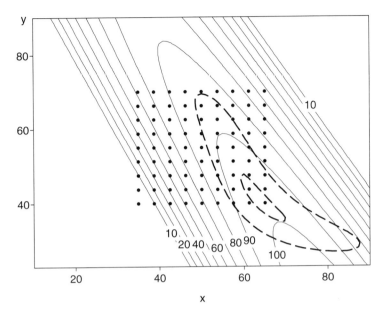

Figure 29. Contours of a quadratic surface fitted to a set of 81 data points. The dashed lines are the level 90 and 100 contours for the true response surface

One way to try to solve the problem is to fit a model which has more parameters. Figure 30 shows the contours for a cubic surface. Such a surface has (in two dimensions) four additional parameters being coefficients of terms in x^3, x^2y, xy^2 and y^3. This provides a better fit to the data.

Figure 31 shows the contours for a non linear response surface. Like the cubic model, this also has 10 parameters. It does have the advantage that it is guaranteed to be positive. In contrast, the cubic fitted response surface takes negative values quite near to the experimental region.

In practice, I recommend fitting non linear models when they take advantage of background information.

The second way of tackling the problem that a quadratic surface is not sufficiently flexible over a wide range is to restrict the range. Since we are most interested in the fitted curve near to the maximum, we should endeavour to restrict the range of the fit to a region near to the maximum.

One possible way of guaranteeing to get silly results should be mentioned here. Suppose we fitted a curve only to the subset of the data which exceed a certain number, say 70. This would bias the fitted response curve upwards. To understand this bias, consider regions where the average response is between 60 and 80. If only data exceeding 70 were considered, then the average response would be more than 70 throughout those regions. Where the true average response was 65, the average non rejected response would be just a little greater than 70. Where the true average response was 75, the average non rejected response would be a little greater than 75, because the lowest possible responses will be rejected.

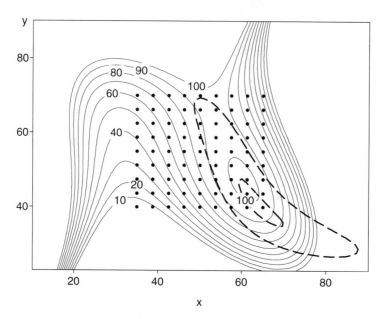

Figure 30. Contours of a cubic surface fitted to the set of 81 data points. The dashed lines are the level 90 and 100 contours for the true response surface

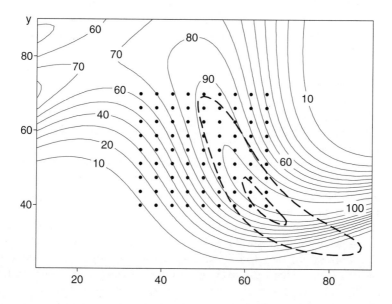

Figure 31. Contours of a non linear model fitted to the set of 81 data points. The dashed lines are the level 90 and 100 contours for the true response surface

DECIDE ON A STRATEGY

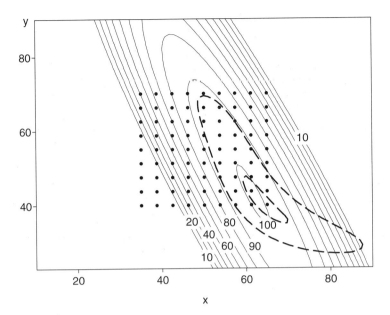

Figure 32. Quadratic surface fitted to the set of 81 data points by weighted least squares such that points with fitted value less than 80 according to the model shown in Figure 29 have very little weight

Figure 32 shows the contours of a quadratic surface fitted with data points being weighted according to the predicted response shown in Figure 29. When the predicted response according to the model displayed in Figure 29 is R, the weight used is the probability that a normal distribution with mean R and standard deviation 7 exceeds 80. This is similar to giving full weight to all points inside the 80 contour on Figure 29 and no weight to points outside that contour, but uses continuously adjustable weights.

The fitted response surface is now quite a good fit over the region of interest. However, it is not as good a fit outside the defined region as the more complicated models shown in Figures 30 and 31.

4.7.7 Lessons to be Drawn

The following points should be borne in mind.

- For response surface methodology, simulation games are useful. Even professional statisticians get very little experience of using response surface methodology on real processes.
- EVOP is not as good as the text books would have us believe.
- The one-hit approach appears to be the most efficient. It also has the greatest risk.

5

Plan a Single Experiment

This chapter deals with planning a single experiment within a series, apart from choosing a design, factors and levels which are dealt with in the next chapter. The tasks discussed here include the following.

- Decide which response variables to measure.
- Try to get more information per unit effort, by increasing the signal, reducing the noise, or sharing the effort and expense with other parties.
- Decide when, where and with what materials to do the experiment, taking into account the likely error variance and the generality of the conclusions.
- Take precautions to avoid or allow for biases.
- Cope with variation due to things other than the factors being deliberately varied: by standardization, randomization or by fitting models which consider these factors.

5.1 DECIDE WHICH RESPONSE VARIABLES TO MEASURE

This section discusses the question of what response variables to measure for a particular experiment. This does not have to be the same as for other experiments within the same programme.

Deciding what to measure is often easy, but at other times it can be a difficult and important problem. Below is a check-list of ideas which might be considered even in those cases when what to measure seems obvious.

Measure quantities which will provide insight into mechanisms and theoretical models. They are more reliable for extrapolating to outside the range of your experimental conditions.

Measure important things, often closely related to total added monetary value, rather than unimportant things. And don't measure everything of any conceivable relevance, intending to work out what is useful later.

Even if data can be recorded automatically so that it seems to have little or no cost, it is valuable to consider what measurements are important. Many organizations which seemed to be inundated with data have failed to measure important things because their confidence that everything was being measured allowed them to fail to consider what measurements would be needed for a particular purpose. However, the recording of extra data, possibly on the grounds that it costs very little to do so, does provide some insurance against changes in the objective or other errors of judgement.

Prefer continuous scales or many-point rankings to two-point classifications. For instance, if you measure characteristics of items produced by a process rather than merely categorizing them as satisfactory or defective, then the direction of a trend may be obvious even though all items are defective or all items are satisfactory.

Don't be constrained by existing or standard tests. Suppose that the accepted test for durability is to boil in water for 24 hours. When doing experiments which require a lot of other work it might be worthwhile to make an assessment of the condition after 1, 2, 4, 8, 24 and 72 hours rather than merely assessing whether a trial passes the standard 24-hour boil test. Alternatively, 24-hour tests at various temperatures might provide a better assessment.

Don't be constrained by tests which are economic for regular use. Testing is only one of the costs of experimentation. If more information per unit of experimental effort can be obtained by spending more effort on sampling or testing, then changes should be made.

Consider measuring timeliness, errors or complaints. Many apparently difficult-to-measure processes such as administrative processes can be monitored by thoughtful choice of measurements, thereby allowing experimental alternatives to be evaluated.

Measure things affected mostly by the single process that you are trying to improve. This will make your picture of what is going on easier to understand.

Ask what matters to the customer, the consumer or the next process. Generally you should try to measure what matters most to these people, the customers of your process.

When trying to identify all of the customers of a process that you are trying to improve, here are some hints.

- The most obvious and generally the most important customers are the ones who are the consumers of the goods and/or services produced.
- Sometimes it is helpful to think of your customer as being a process, rather than identifying a large number of individual people and listing them as separate customers. In such cases the customer's requirement can often be thought of as being that your output causes no problems for the customer process.

    For instance, a company mining iron ore might identify its primary customer as the process of making iron in a blast furnace and consider whether the properties of the ore are useful for that process, rather than listing all of the companies that purchase its iron ore and considering their specified requirements.
- Sometimes the people who pay you are not the same as the consumers.

    — There may be an intermediate level of sales people.
    — The consumers may be children.
    — Costs may generally be covered by government or insurance.

    In cases like this it is necessary to satisfy both the consumers and the payers.
- There are often other parties who have an interest in the outputs of your business. The owners of the business would like it to be profitable, the local authority would

PLAN A SINGLE EXPERIMENT

like it to minimize its use of services such as street-cleaning. Other residents of your neighbourhood might be concerned about your emissions of noise or smoke. It can sometimes be useful to regard all of these as customers.

Sometimes these classes of customers are referred to as 'stakeholders' to distinguish them from customers who are interested in your main outputs.

- The people who will repair or clean your product, or who will provide subsidiary services are also customers because what you do affects their processes.

It is common for a company to assess incoming goods to check that they will perform adequately in the company's processes. Normally the company specifies that the incoming goods must satisfy some well-specified conditions. However a supplier would do well to remember that the real requirement is that the goods perform adequately. There would be no point delivering goods which satisfied the well-specified conditions but did not perform adequately, because the receiving company would quickly change the specified conditions.

Ask what distinguishes good from bad. There are generally many ways in which products or services can be bad, and each of these can be the basis of a quantity to be measured. Counting the total number of defectives without distinction is virtually never an efficient way to proceed.

Ensure that measurements are feasible and reliable. For unfamiliar measurement techniques, check them before starting the experiment.

When the purpose of experimentation is to improve the robustness of some product or process then it is frequently useful to construct a performance statistic from the results of more than one run. For instance, Pignatiello and Ramberg (1985) discuss an experiment intended to make the height of a particular type of truck spring less sensitive to the temperature of the oil bath used to quench the spring after heat treatment. Sensible performance statistics are

- the height of a spring made at low quench temperature minus height of a spring made at high quench temperature,
- the standard deviation of heights of springs made at variable quench temperature, and
- the coefficient of variation of heights of springs made at variable quench temperature.

Any of these would be of some use in this application. They are all quite different from merely analysing the mean spring height.

Some questions to ask yourself when choosing a performance statistic are as follows.

- Would it be appropriate to transform the scale of measurement?
- Is there a performance measure which would be little affected by adjustment of the process mean? For instance, for some electronic circuitry, the signal-to-noise ratio is a useful performance statistic. If experiments or design efforts are used to improve the signal-to-noise ratio then that improvement is generally retained when the average level of a signal is modified.
- Can an operating window be defined so that its size is a good measure of robustness of a product or process? For instance, for a paper feeder for a photocopier, the range of pressures on a feed roller or the range of paper types which give neither

misfeeds nor multifeeds under well-defined conditions is a good definition of an operating window. Making the process more robust may be interpreted as enlarging the operating window.

Although you should think about what you would like to measure, you also need to ensure that the effort of measuring does not unreasonably constrain the number of trials that you are able to conduct.

Show a bias towards action. When the effort of measuring experimental outcomes is great compared to the effort of conducting trials, do not let the perceived difficulties worry you into a state of inactivity. It may be better to quickly conduct a few trials using informal measurement procedures, than to procrastinate and thereby delay any benefits to be gained from experimental runs.

Subjective assessments are quite permissible. They are quick and cheap. They are also likely to consider all aspects of the outcome. This can be important when it is difficult to get objective measures of all aspects of outcomes. However, where subjective assessments are used it is important to ensure that the order of assessment is balanced across treatments. Try to help the assessor(s) to maintain consistency by providing reference samples or photographs. If the important properties are essentially subjective (e.g. taste) then face this fact. Data can be quite valuable, even if it consists of subjective judgements.

You don't need to measure the same thing all the time in a sequence of experiments. For instance, it may be appropriate to measure outcomes in a quick or cheap way during some exploratory trials, while finding out a range of conditions which make a process work reasonably well; and to make a small number of more expensive or slower measurements at a later stage of experimentation.

5.2 SEARCH FOR EFFICIENCIES

In a sense, the only efficient experiment is one with only one trial. You try only a single set of conditions and they happen to give excellent results. All other experiments will, in hindsight, look inefficient compared to this perfect experiment. The review of 32 student projects by Hunter (1977) found that it is common for people to discount the extent of what they have learnt, because with hindsight they feel that they should have expected much of what they observed.

Such efficiency is not generally possible, because we only experiment when we lack knowledge. And even this apparently ideal experiment has the undesirable feature that you would have no basis for being confident about whether or not even better results were possible.

5.2.1 Regarding Experiments as a Competition between Signal and Noise

An experiment is like a competition between the effects you are interested in (signal) and the effects you are not interested in (noise). It differs from many other competitions in that you are able to influence the rules. And, even less like games people play for recreation, you can change the rules between rounds of competition if you don't like the rules that you tried first.

PLAN A SINGLE EXPERIMENT

Some ways of increasing the effects that you are interested in are to

- use more extreme conditions than will be used in later practice, and;
- use particularly sensitive measuring equipment.

Some ways of reducing the effects that you are not interested in are to

- reduce variation by blocking, or use of covariates;
- average out variation by replication;
- standardize operating conditions;
- provide aids for consistency of measurement or rating;
- homogenize raw materials before taking samples for the various runs;
- calibrate measuring equipment frequently, and;
- use blind or double-blind trials (if appropriate).

In this section we will be concerned with getting as much information as possible per unit of effort. This means both maximizing the amount of information and minimizing the amount of effort. We will be interested in questions like the following.

- Can the quantity of interest be made larger so that it can be more readily detected?
- Can the amount of noise be reduced, possibly by use of blocking or covariates?
- Are there other ways of improving the ratio of useful information to effort?
- Can our suppliers, customers or competitors be induced to participate in this experiment?

There are three types of strategies, namely reducing the noise, increasing the signal, and increasing the information-to-effort ratio. They will be discussed in turn.

5.2.2 Understanding the Noise in Experiments

When you conduct experimental trials you do not generally expect the results to be perfectly reproducible. The causes of this lack of reproducibility are referred to as 'noise'.

Example: Fruit tree spraying

One experiment, that I remember with some feeling of guilt, involved counting the numbers of insects at the various stages of their life cycles on each of several leaves on each of several limbs on each of several fruit trees at each of several times during a season. The purpose of the experiment was to compare two different regimes for spraying the fruit trees. There were four sets of six fruit trees used in the experiment. Two sets were sprayed according to one spraying regime and two sets were sprayed according to the second spraying regime.

I was in a position to anticipate that the effort required to conduct this experiment would be large compared to the useful information likely to be obtained. I felt guilty because I failed to discourage the experimenter from wasting six months of his working life.

The key to anticipating that this experiment would be of little value is to think about the barriers to its reproducibility. The patterns of insect infestation and the effects of spraying would be likely to be quite different in another season. In one season, spraying at a particular time of year might be very effective. In a second season, spraying at the same time of year might be reduced in effectiveness by the weather. In a third season, spraying might interfere less with the insects that spoil the fruit than with other insects which are their predators. Yet the experiment does not average over seasons and provides no information about the amount of season-to-season variation.

The large amount of data to be collected tended to give the feeling that the experiment would be more reliable than it actually was. The six trees in each set were subject to a similar pattern of insect infestation because they were near to one another. They also experienced very similar rainfall, soil fertility, drainage, sunshine and levels of attack from soil-borne pests. Similarly, the various limbs on the trees, the various leaves on each limb and the many occasions of counting do not make the experiment much more reliable. There is a lot of variation between tree, limbs, leaves and occasions of counting, but these sources of variation are not major barriers to reproducibility of the experimental outcome.

5.2.3 Reducing the Noise

The first step in reducing noise is to identify the things that cause it.

Factors not being investigated should be kept as uniform as possible during the conduct of an experiment. It is generally advisable to complete an experiment quickly so that aspects of the environment and equipment not explicitly considered are kept reasonably constant.

Quantifying the amount of noise from each source is often the next step in solving problems about noise.

Example: Quality of cement

The manager of a cement plant thinks that the variation in the quality of its product is partly due to:

- variation in raw materials between various sources;
- variation in raw materials caused by age and stockpiling practices;
- factors which are readily controllable during the reduction process, and;
- storage after manufacture.

In addition, there are several sources of variation associated with sampling and testing the raw material and the cement.

Experiments performed in ignorance of some of the sources of variation would be likely to be misleading. For instance, the effect of optimizing some readily controllable factors using only raw materials from a single source might look promising, but turn out to be far from optimal for raw materials from other sources.

Example: Heat treatment of papayas

Chan et al. (1996) compare the effects of various heat treatments on papayas. The response variable measured was the activity of the ethylene forming enzyme system, which is also called ACC Oxidase. This is measured on discs cut from fruit.

PLAN A SINGLE EXPERIMENT

John Maindonald's crucial contribution to this research was to point out that single fruits could be monitored over time, whereas the other researchers 'had previously used excised tissue from several fruits and pooled or mixed the excised discs. For individual experiments discs were then randomly selected from the pool.' This enabled the noise in the experimental results to be reduced, because much of the noise was previously due to the large amount of variation from one piece of fruit to another.

Example: Variation in election results

Political commentators generally attempt to estimate the final result of elections after about a quarter of the results have been reported. The main difficulty in this estimation task is that booth-to-booth variation is very important within an electorate. When a quarter of the results have been reported, generally complete counts have only been reported for a fraction of the polling booths.

The reliability of the current percentage of votes for given candidates depends on the consistency of the voting pattern across booths and the number of booths which have been counted. The percentage of votes counted is not a good indicator of reliability. Estimates of the final results can often be made more precise by taking into account the pattern of booth-to-booth variation in previous elections.

5.2.4 Replication

One way to reduce the influence of noise is simply to take replicates. You have to be careful about what is meant by replicates and also about economics.

Roughly speaking, a replicate is repeating a run for the same levels of all of the experimental factors. However, the term 'replicate' is often a source of confusion. It is my belief that the confusion arises largely because people tend to concentrate on the things that are the same for replicates rather than thinking about the things that can be different between replicates.

A two-step procedure[1] for deciding whether supposed replicates are really replicates is as follows.

1. List the sources of variation which affect runs made at the different experimental conditions being investigated.
2. Check that those sources of variation are allowed to vary between the supposed replicates.

For instance, suppose that one of the factors in an experiment is the way of setting up a machine. Amongst the sources of variation which affect runs made at the different experimental conditions being investigated are the vagaries of the setting-up process which may include torques with which nuts are tightened, alignment of components

---

[1] Teresa Dickinson and I formulated this procedure based on our experiences of trying to explain definitions of the term 'replicate' in training courses. When people made mistakes about whether or not supposed replicates were really replicates we found that the problem was almost always in their failure to identify a source of variation. We thought about a definition of 'replicate' as a procedure for checking whether supposed replicates were really replicates, and tried to make this procedure more reliable.

and calibration of feedback mechanisms. Repeating a run without repeating the processes required to set up the machine would not be a 'replicate'.

We are using the term 'replicate' as an abbreviation for a 'replicate experimental run'. Terminology such as 'replicate samples', 'replicate measurements' and 'replicate determinations' is sometimes used. When I hear such terms being used, I hope that they will tell me how many of the processes of sample extraction, sample preparation, set-up of measurement equipment, measurement operation and generation of the number that is the result have been repeated. Usually, despite the best efforts of people who try to standardize terminology, my hope is not justified. So I ask the question that I would encourage you all to ask in this type of situation: 'Which sample processing steps have been done separately for the things described as "replicates"?'

Repeating part of an experimental protocol should not be described as producing a 'replicate experimental run', but it can be extremely valuable. It will reduce those components of noise which are replicated and will thereby improve precision. The expected improvement in precision can be calculated provided that the variance of the components of noise have been estimated. The expected improvement can be balanced against the increase in effort and cost in order to optimize the amount of replication.

5.2.5 Randomization

Randomization means conducting some parts of the experimental procedure in a random order. It might mean taking samples of raw material in a random order, assigning individuals to treatments at random, conducting the trials in a random order or testing the results in a random order. Randomization does not decrease the amount of noise. However, it does ensure that the average influence of the noise on the comparisons of interest is lessened.

One alternative to randomization is conducting trials in a systematic order. Any noise which drifts slowly will be likely to have a major influence on such an experiment. For instance, suppose that a 3×3 experiment with factors A and B were conducted in systematic order with the first three trials being for the low level of factor A and the last three trials being for the high level of factor A. Any slowly varying source of noise would be much more likely to have a substantial influence on the estimate of the effect of factor A than would be the case if the order of runs were randomized.

Even worse, imagine doing a similar 3×3 experiment on nine sheep in an enclosure, applying the low level of factor A to the three sheep which you are able to catch most quickly and applying the high level of factor A to the three sheep which evade capture for longest.

Example: Interpretation of road signs

In 1976, soon after I had started working at the Australian Road Research Board with a Ph.D. in statistics but with little practical experience, I was asked to analyse some data from an experiment on the interpretation of road signs.

Some new road signs were being tested. They used diagrams and symbols but did not include any written words. The assumption behind the experiment was that people with very little knowledge of English would be likely to find the new signs easier to

PLAN A SINGLE EXPERIMENT

interpret than the existing signs which relied on words. The experiment was designed to find out which signs were most readily interpreted by the majority of drivers, whether or not they had a satisfactory knowledge of English.

I thought that one feature of the experimental design was unsatisfactory. The candidate road signs were presented in exactly the same order to all of the people used as experimental subjects. I was concerned that this might bias the results. For instance, if two candidate signs were similar and both were meant to be interpreted as, say, 'give way', then I would expect the sign presented first of these two to be less well interpreted than the sign presented second.

I felt very uncomfortable about being asked to analyse the data from this experiment, because I thought that I would be pressured to suppress my misgivings about the reliability of conclusions which might be drawn by routine statistical analysis. I escaped from the meeting without promising to analyse the data, but I felt very uncomfortable about not being able to satisfy the requirements of one of the very senior researchers, Jim Bryant.

The next day during morning tea, Jim Bryant told one of his colleagues in a way that I was sure that I was meant to overhear that I had told him '...to tell his own lies!'

5.2.6 Blocking

Blocks are sets of trials which would be expected to be more similar than trials in different blocks, if it were not for different treatments being applied. For instance, trials made using a single batch of raw material might be treated as a block.

Experiments which are blocked are generally analysed using only within-block comparisons. This wastes a small amount of information, but the practice of blocking usually improves experimental precision because treatments are compared with less noise.

Blocking is not appropriate if there is expected be an interaction between treatments and blocks – that is a tendency for treatments to have different effects in different blocks.

When considering options for blocking, remember the principle of minimizing variation within blocks. Do not make blocks consist of large numbers of trials if this will increase the residual error variance. If measuring equipment must be recalibrated from time to time, ensure that all measurements made on a single block are made within the same calibration.

In some circumstances it is difficult to find an experimental design which can be blocked as well as having other desirable properties. For instance, a metal rolling mill might start with pieces of metal which can be expected to be very similar within groups of three pieces, but factors being investigated might each have two levels. This makes selection of a small, balanced design rather difficult. In such circumstances I suggest that blocking be given higher priority than balance.

5.2.7 Covariates

Covariates are extra variables which help to explain variation in experimental results. For instance:

- ambient temperature, humidity or quality of incoming material might help to explain the variability in some aspect of the quality of some product;
- the quality of a product produced in the standard way may be a useful covariate for experimental production runs;
- in agricultural trials, the unavoidable variation in soil fertility might be partially explained by the yield of something grown the season before an experimental crop is grown.

The model that is fitted might be

$$Y_i = t_j + \beta X_i + \text{error}$$

where Y_i is the experimental yield in the ith run in which treatment j has been applied, β is a constant to be estimated, the t_j are treatment effects to be estimated and X_i is the covariate.

When analysing data involving covariates it may be helpful to move the origin of measurement of the covariate so that the treatment effects are easier to interpret. The covariate may be measured relative to a natural origin, so that the treatment effects can be interpreted as applying when the covariate takes that value. Alternatively, the mean value of the covariate may be subtracted, so that the treatment effects can be interpreted as applying when the covariate takes its average value. In both cases the estimate of β tells us how much change in response is associated with unit change in the covariate.

5.2.8 Increasing the Signal

Sometimes it is possible to make the effect of interest larger, thereby increasing the efficiency of experiments. With factors which are continuously variable, using more widely spaced levels is a way of increasing the signal.

With experiments to estimate the variance of sampling errors, it is easy to use smaller samples than would be used routinely. This makes the sampling error variance larger, because sampling variance varies inversely with sample mass. Therefore sampling errors become large relative to measurement errors, making it easier to estimate the amount of sampling error. This technique was discussed in the example on sugar cane sampling error on page 76. It has also been used to estimate the sampling error for iron ores, see Robinson and Holmes (1995).

Example: Bias of a flap sampler

Hammer-milled sugar cane is sometimes sampled using a flap sampler like that indicated in Figure 1. Such a sampler is likely to be biased because the right hand side of the stream of incoming material is diverted to the sample chute for longer than is the material on the left hand side of the stream.

The bias of the sampler might be acceptable because the cane has just been mixed very thoroughly by being hammer-milled. The bias of the sampler could be estimated directly by comparing samples of the entire stream of incoming material with samples taken by the flap sampler.

An alternative requiring less precise experimentation is to look at the characteristics of the incoming stream of material on the left, in the middle and on the right. We are

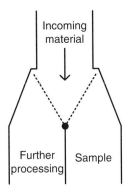

Figure 1. Diagram of a flap sampler. A sample is taken as the movable flap (shown in two different positions by dotted lines) moves from right to left and back to the right

primarily interested in seeing whether the left hand side of the stream is different from the right. The geometry of the motion of the flap sampler can then be used to estimate the sample bias from the left–right segregation of the incoming material. Because the left–right segregation of the incoming material will be larger than the bias of the flap sampler it will be easier to measure it.

5.2.9 Increasing the Information-to-Effort Ratio

In the previous two subsections we have been discussing ways of improving the signal-to-noise ratio. We now look at other ways of getting more useful information relative to the amount of effort or money expended.

Agricultural trials of many types can be conducted only once per year. The time cost of a single small experiment is therefore the same as that of a large complex experiment. This encourages agricultural researchers to do complicated trials.

In most other areas of experimentation a sequence of small, simple trials is most cost-effective.

5.2.10 Split-Plot Designs

Sometimes it is convenient to apply some factors to large experimental units and to apply other factors to smaller experimental units. When this is done, the large experimental units are referred to as 'main plots', the smaller experimental units are referred to as 'subplots' and the design is described as being a 'split-plot' design.

An agricultural situation where a split-plot design might be used is with factors such as irrigation treatments, which can only be applied to large areas of land, and plant varieties which can easily be varied from one small area to another.

People conducting industrial experiments often incorrectly ignore the split-plot nature of their experiments. For instance, in the experiment on a wave soldering machine described on page 81 all runs with old solder flux were conducted before all of the runs with new solder flux.

When experiments are conducted as split-plot designs, treatment comparisons which can be made using measurements on subplots are much more precisely estimated than are treatment comparisons which can only be made using main plots. This is both because there are many more subplots and because the soil or experimental material is generally more uniform between subplots within the same main plot than between subplots in different main plots. This may or may not be an issue. It is not essential to estimate the effects of all factors with equal precision.

The analysis of split-plot designs is complicated by the fact that two different components of noise must be estimated. It is also necessary to be careful about which treatment effects have which precisions.

5.2.11 Sharing the Cost

It is often possible to reduce the cost of experiments by sharing it with other parties who are also interested in the results.

- Experiments near to the start of your processes are likely to be of interest to your suppliers.
- Experiments near to the end of your processes are likely to be of interest to your customers.
- Experiments studying sampling and testing precision are likely to be of interest to your competitors as well as your customers and suppliers.

5.3 DECIDE WHEN, WHERE AND WITH WHAT MATERIALS

For many experiments, there are difficult decisions to be made about when to conduct the experiment, where to conduct it, or what materials to use. You must try to reduce error variance and expense, subject to the requirement of drawing inferences about the population of interest, not a restricted population.

- For industrial trials:
  - Be adventurous with pilot plants. Remember that you do not have to provide a final solution, but you should aim to provide guidance to the people who will experiment at full scale.
  - When experimenting at full scale, carefully weigh the likely value of information against the cost of obtaining it.
- For medical and animal trials, consider using related animals or using the same animals at different times for different treatments.
- For agricultural trials:
  - Blocks should be as uniform as possible. They are likely to be nearly square, but consider all information about the site.
  - Plots should be long and thin.
  - Check that plots are large enough to enable treatments to be conveniently applied to plots. Plots should generally be as small as possible subject to this constraint.
  - Consider possible interactions between plots due to competition below or above ground, insects, etc. Can competition be reduced or allowed for?

PLAN A SINGLE EXPERIMENT 125

Following on from the previous chapter, we may want to reduce noise or variation, but we also want to keep the experiment relevant. The two issues of reducing error variance and ensuring general applicability are often in conflict with one another. Methods of reducing variability often make an experiment look as if it might not be relevant to run-of-the-mill production. Resolution of this possible conflict must be done on a case by case basis.

The siting of an experiment means deciding on which piece of land to do an agricultural field trial. Decisions of similar importance are deciding which piece of equipment or raw material to use for experimental runs, choosing a laboratory to do the testing, deciding whether to use a single operator or many operators, and deciding which times of day are to be used for trials. Such decisions may be very important, but they are often neglected.

The crucial issues are considered below.

5.3.1 Can the Error Variance be Reduced by Siting the Experiment in a Particular Way?

Generally, the variation between experimental plots for agricultural field trials is minimized by using long, narrow plots, but local conditions must be taken into account. Variations in fertility due to previous land use, differences in drainage, differences in soil type and effects of plant competition all need to be considered. The variability of soil may be understood by thinking about events in the soil's history and may be quantified by monitoring the error variance of experiments conducted using various types of soil.

If either a new, high-capacity machine or an old, low-capacity machine may be used for experimental manufacturing then the old machine tends to be selected because the opportunity cost of using it is less. However any difference in uniformity of output ought also to be considered. An experiment using less variable equipment may be smaller for a given precision.

Using a single operator would reduce noise, but using many operators would give wider applicability to the results.

When testing errors are a substantial fraction of total variation in an experiment, more precise measuring equipment should be considered.

To do experiments near the middle of the day – when equipment and people have got into uniform work rhythms, atmospheric temperature and humidity are not changing rapidly, and there are no end effects due to the ends of the working day – may reduce variation for some types of experiments.

5.3.2 Is it Reasonable to Assume that Results at the Site Chosen Apply in General?

There is no point finding out things which cannot be applied. Although the siting of an experiment should be done with an eye to reducing variability, the siting should also consider the ease with which the results can be generalized.

Meticulous weeding of plots for agricultural field trials might reduce variability, but might also make the results less relevant because the plant varieties or management regimes which produce best yields given meticulous weeding are likely to be different from the plant varieties or management regimes which produce best yields given realistic amounts of weeding.

Using more precise measuring equipment than usual generally does not interfere with the applicability of results.

5.3.3 Making the Compromise

Consider some examples of decisions – about when, where, and with what materials – which might be made. In each of these cases there is a decision to be made which amounts to finding the best compromise between reducing error variance and ensuring general applicability. My recommendation is that general applicability must be checked at some stage during a programme of experimentation. Apart from this stage, it is probably best to give reducing error variance priority over ensuring general applicability.

- Using distilled water for chemical experiments is good from the point of view of reducing error variance, but may be bad from the point of view of ensuring that the results are generally applicable. It is generally desirable to investigate whether cheaper water would be satisfactory at some time during the development of a commercial chemical process.
- An experiment on a mineral processing plant could be much more precise about its conclusions if it was conducted using blended ore or conducted at a time when reasonably uniform ore was being mined. However, drawing inferences about what would happen with low grade ore would be unreliable.
- Sampling and testing can be made more consistent by using a single operator. However, if conclusions about sampling and testing precision needed to be applied to all operators then the experiment would not be relevant.
- Where there are several manufacturing lines or several operators, comparisons can be made more precise by doing the entire experiment using a single line and a single operator. However, some conclusions might not be applicable to all lines or to all operators. Similarly, changes to operating procedures might be compared most precisely by experimenting during a night shift, because there are fewer interruptions.

5.3.4 Scaling Up

It is common for experiments to be conducted first on a laboratory scale, then on a pilot plant scale and for full-scale trials to be conducted later. Deciding which questions to investigate at which scale is a particular case of 'when, where, and with what materials'. Understanding which aspects of small-scale experiments are likely to need modification for full-scale production is very important for the cost-efficiency of research. I have some suggestions.

- If you have very little background knowledge then it is very important to do some work at the laboratory scale.
- Investigation at laboratory scale is generally much cheaper than investigation at larger scales. Experimentation at the laboratory scale should be sufficiently extensive that if initial trials at the pilot plant scale or at full scale do not give

PLAN A SINGLE EXPERIMENT

the desired results then the knowledge obtained from laboratory scale experiments will give useful ideas as to promising ways of modifying the process.

- A research laboratory should consider the scaling-up process as applying to many research projects, monitoring the successes and failures of this process to see how it can be improved. If there are aspects of full-scale performance which are not well predicted by laboratory scale experiments then you should plan to spend more resources on pilot scale and full-scale experiments.
- While working with small-scale experiments, anticipated changes to the conditions for full-scale production should be investigated or used routinely. For instance, use commercial grade reactants rather than laboratory grade. Allow for uneven mixing and uneven temperature distribution. Consider the effects of using dirty equipment.
- As a general rule, if the effect of a factor shows little interaction with the levels of other factors then scaling-up is unlikely to be a problem.

5.4 TAKE PRECAUTIONS TO AVOID OR ALLOW FOR BIASES

Bias is a tendency for estimates of a characteristic to be persistently higher or persistently lower than the true value. If a bias remained constant during an experiment then it would not usually be a problem, because the relative effects of the various factors would be not be influenced by the bias. However biases are not perfectly consistent. Most biases drift with the passage of time or vary with some aspects of the things being tested.

Sometimes biases in an experiment would mean that even an experiment of infinite size would not be adequate to give you the information that you want. This section discusses how to avoid biases or to allow for them, considering questions such as the following.

- Are the sampling procedures unbiased? Do all particles have the same chance of becoming part of a sample?
- Are testing procedures well proven, calibrated and precise?
- Are criteria for subjective judgements defined as clearly as possible?
- Is it possible that conscious or subconscious bias could arise? Should information be withheld from subjects or experimental staff? Would disclosure of information and extensive explanations increase cooperation?

5.4.1 Avoiding Sampling Bias

Taking and preparing samples are possible sources of bias. Procedures for sampling gases and liquids are likely to be unbiased, provided that there is no segregation due to differences in density. And remember that differences in temperature can easily give rise to differences in density. However, sampling of solids, suspensions and mixtures of solid particles and liquids are generally prone to bias. We all need to check our procedures.

An illustration of one of the difficulties of sampling is based on a suggestion of Scheaffer (1997). Your task is to estimate the average area of the shapes in Figure 2 by selecting a set of, say, six shapes and averaging their areas. I recommend that you photocopy page 128 and select six shapes before reading on.

Figure 2. A set of rectangular shapes. Readers are invited to try to choose a representative sample of, say, six of these shapes and to calculate the average area of the shapes selected

PLAN A SINGLE EXPERIMENT

Many people are aware that there is a tendency to be more likely to select large shapes than to select small shapes, but it is not easy to overcome this bias. Some reasonable strategies are the following.

- Number the shapes. Use a random number generator to select a sample of numbers. Choose the shapes corresponding to the selected numbers.
- Shade in one square from each shape. Then throw darts, choosing shapes when a dart hits a shaded square.
- Draw a line across the page and choose the shapes whose geometric centres are nearest to the line.

For readers wanting to compare their samples of shapes with the population mean, here is some relevant information. There are a total of 100 shapes on page 128. The numbers of shapes of various areas are shown in this table.

Area	1	2	3	4	5	6	7	8	9	10	12	14	15	16	18	20	21	24	25	27	30	40
Number	22	15	10	10	3	5	2	4	4	4	4	1	1	2	2	2	1	3	1	1	2	1

The average area of the 100 shapes is 7.19 and the standard deviation of the areas is 7.98.

Many sampling procedures are likely to be biased because of simple issues which make some parts of a lot less likely to be included in a sample. If the full extent of a stream of material is not equally sampled, then samples cannot be expected to be representative. There can never be a guarantee that the unsampled part of the stream is similar to the sampled part of the stream. Some other issues in avoiding sampling bias are as follows.

- For slurries, this means sampling at a point where the material is falling, as for solid particulate materials.
- If a scraper is used to remove fine, damp material from a belt, then it should be designed so that the scrapings are also sampled.
- Use of dust extractors will reduce the probability of ultrafine material becoming part of the sample to be tested. This may be desirable for health reasons, but the extent of the bias to sampling should be estimated.
- Sampling from stationary situations such as trucks and stockpiles is very difficult to do properly. Can you ensure that the particles on the floor of the truck or at the base of the stockpile have the same probability of being sampled as particles which are more easily accessible?
- Removal of large particles from a sample (because a comminution device is designed for particles of less than a specified size) should be expected to lead to bias. An estimate of the bias can be obtained as the proportion of the sample removed times the difference in average grade between the material removed and the remainder. Such an estimate of bias should not be used to correct the results obtained but can be used to decide whether to bother about reducing the bias by sampling more correctly.
- Loss of sample and contamination should be avoided. In particular, inconveniently heavy containers of sample should not be made lighter by simply tipping some sample out.

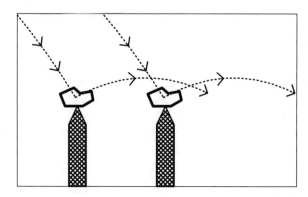

Figure 3. Possible trajectories of two particles separated by a distance of precisely the cutter aperture

- Sample cutters which take some particles from a falling stream of material should have a reasonably large aperture and should not travel too fast. Otherwise they tend to undersample large particles because large particles are more likely to bounce off the cutter blades and miss being sampled than are small particles. Figure 3 illustrates two particles of the same shape which bounce off a pair of cutter blades. They are shown bouncing sufficiently far that neither of them ends up between the cutter blades, so neither of them ends up being in the sample. If the distance between the cutter blades had been larger then one of the particles would have ended up being in the sample.
  The currently accepted rules, due to Pierre Gy and explained in Gy (1979) and Pitard (1989), are that the cutter aperture should be at least three times the minimal topsize of the material being sampled and the speed of the cutter should be not more than 0.6 metres per second.
- Sample cutters should have the correct geometrical properties for taking an unbiased cut of the entire stream of material. They must intercept the entire stream of particles. They must move at constant speed. The cutter blades should be sharp and straight. The cutter opening should be constant and the cutter should have sufficient capacity to hold the sample.

It is often worthwhile to check the extraction ratio. Given a cutter aperture of a metres, a cutter speed of v_c metres per second, and ore flow rate of G kilograms per second, a cutter should extract a/v_c seconds of ore flow. The mass of sample which should be expected is therefore Ga/v_c kilograms. The extraction ratio is the average of the amount of sample obtained divided by the mass of sample expected. The extraction ratio should be near unity. For it to exceed unity suggests that your samples are surprisingly large or that your measurements or calculations are inaccurate. If the extraction ratio is less than unity then some of the particles which should have become part of your samples have been lost for some reason. The problem should be investigated.

5.4.2 Avoiding Measurement Bias

This subsection is about avoiding bias when measuring. I am taking a rather broad definition of measurement bias. Some of what I have to say in this section might be regarded by other people as commenting on what should be measured.

Example: Weighing shipments of iron ore

Several years ago I was told about some comparisons of the mass of shipments of iron ore. For ships loaded in Australia and unloaded in Japan there was reasonably good agreement between the assessments of the mass made at the two ends of the journey. There was little, if any, bias and the precision seemed consistent with the precision expected of the draft survey methods being employed. However, for ships loaded in Australia and unloaded in Europe there was a small bias. This bias was obvious when large numbers of shipments were compared, with the measured mass being generally larger at the Australian end of the journey.

On loading ships in Australia, the mass of cargo was assessed by doing a draft survey of the fully-laden vessel and subtracting the registered mass of the vessel, adjusted for the amounts of water and diesel on board. In both Japan and Europe, the mass of cargo was assessed by doing draft surveys before and after unloading and comparing the results. Payment for the cargo is based on the mass and grade of the cargo as measured at the unloading port, for both Japan and Europe.

The fee paid for carriage of ore was based on the Australian assessment of mass for shipments to Europe, but was based on the Japanese assessment of mass for shipments to Japan. Hence, the masters of ships taking iron ore to Europe had a financial incentive to slightly understate the amounts of water and diesel they had on board, in order that the amount of ore be overestimated.

Example: Soil with sugar cane

As with other heavy industries, maintenance is a major expense for sugar mills. One cause of wear of the equipment in sugar mills is that some soil is attached to the sugar cane as it is delivered to mills. The amount of soil is affected by the height above the average ground level at which the cane is cut. There is an economic incentive to the cane growers to cut the cane lower because this increases the tonnage of cane harvested.

Sugar mills wished to provide an economic incentive for farmers to deliver cleaner cane. However, accurate measurement of the amount of soil included with loads of cane would be very expensive.

One mill decided to use a subjective visual assessment of the cleanliness of the cane, and found a creative way around the problem that the assessment was likely to be biased. The mill structured a bonus and penalty scheme so that some growers were paid a bonus for delivering cleaner than average cane, while other growers were penalized for delivering dirtier than average cane. The total of the bonuses was guaranteed to precisely match the total of the penalties over a negotiated time frame. Therefore neither the mill nor the farmers would be affected by the average bias of the assessments.

The mill was able to provide an incentive to cane growers to deliver clean cane without the expense of installing an accurate sampling and measurement system.

Calibration

Attempts to calibrate measuring systems are made if it is thought that the measuring systems might drift. However, all calibration systems are imperfect so calibration should itself be regarded as a source of noise in measurements. The minimum amount of concern with calibration during experiments should be to note when relevant measuring equipment was calibrated.

A situation to avoid would be doing all of the measurements for one level of a factor before recalibrating a measuring device, then doing all of the measurements for another level of that factor. If this happened then any calibration would be confounded with the effect of that factor.

5.4.3 Blinds and Double-Blinds

One aspect of avoiding bias (and also avoiding the appearance of bias) is to ensure that human subjects – people delivering treatments and people rating outcomes – do not know which treatments have been applied.

An experiment which involves human subjects is described as 'blind' if the subjects are not given information about treatment identification, so that their prior opinions, beliefs and prejudices cannot affect the results. It is described as 'double-blind' if information about treatment identification is also withheld from other people involved in the experimental protocol. See Gore and Altman (1982, pages 51–53) for discussion of this topic in medical contexts.

A crucial consideration in deciding whether or not to randomize an experiment is whether in the absence of randomization alternative hypotheses might explain some aspects of the results.

The same principle tells us when to use blinds or double-blinds. We should use blinds or double-blinds if, in the absence of such precautions, alternative hypotheses might explain some aspects of the results. Suppose a greengrocer had asked some of his customers to rate the tastiness of two varieties of apples, good old reliable Granny Smith and new genetically-modified Nerd Special, and found that most of them preferred Granny Smith. Would you regard this information as reliable? Would your opinion change if the apples presented to customers were not identified? Would your opinion change if the greengrocer was also kept in ignorance of the identity of portions of apple, which were peeled and randomly labelled 'A' and 'B' by a person who did not communicate with either the subjects or the greengrocer?

Sometimes, even double-blind experiments do not disguise the treatments adequately in order that bias be avoided. In many circumstances, subjects or investigators who have not been told which treatments are being used on which subjects will tend to make their own inferences about the treatments. See Philipson and Desimone (1997) for a discussion of this problem.

5.4.4 Examples

- If wines are to be assessed on the basis of taste, appearance and smell then it is important that the tasters not be informed of the identity of the samples being tested.

PLAN A SINGLE EXPERIMENT

- In trials of the possible effect of vitamin C on the common cold, it is correct practice to provide sugar tablets (placebos) for volunteers not receiving vitamin C tablets and to not tell the medical staff providing the tablets which bottles contain vitamin C. A volunteer with a positive attitude is less likely to catch a cold and the attitude of medical staff tends to affect the attitudes of patients.
- Gould (1996) discusses how the measurement of skulls has sometimes been affected by the measurers' knowledge of the race of those skulls together with their preconceived opinions.
- Tanur et al. (1972, pages 2–13) discuss experiments to investigate the effectiveness of the Salk polio vaccine. In some counties double-blind trials were conducted. In other counties, volunteers in grade 2 were inoculated and compared to non volunteers in grade 2 and the totality of grades 1 and 3. The results illustrate that there was a volunteer bias. Data on the variation of polio infection rates from year to year also illustrate the danger of using historical controls.
- Some experiments which were conducted at the Hawthorne Plant of the Western Electric Company, in Cicero, Illinois, USA between 1927 and 1932 started off investigating the effects of factors such as level of illumination, wages, timing of rest periods, length of working weeks, temperature and humidity. The greatest increases in average work output were found to have been caused by changes in attitudes which happened because the company had shown an interest in the workers whose output was being monitored. This has been called the 'Hawthorne effect'. It is a hypothesis which might explain trends in many experiments on people if blinds and double-blinds are not used.

5.5 COPE WITH OTHER SOURCES OF VARIATION

In particular, randomize the way of conducting the experiment unless this is not feasible or is particularly expensive. If intending not to randomize then ask the question 'will failure to randomize be a barrier to acceptance of the conclusions of the experiment?'

There are three strategies for coping with sources of variation which are not included in an experiment.

- Effort can be made to standardize them, at least within blocks. (See page 121 for a discussion of blocking.)
- Randomization can be used to reduce their effects.
- Their effects can be removed by statistical analysis after the experiment has been conducted.

5.5.1 Standardization

Example: weeds in plots

A colleague, Bronwyn Harch, was asked to analyse some data from an experiment looking at the effects of various types of sewerage sludge on the balance of populations of microbes in soil. The experimental plots were cement tubs, placed in holes in the ground so that the level of soil in the tubs was the same as the level of soil outside the tubs.

She found that the differences between treatments were of similar magnitude to the differences between plots which had been treated with the same type of sludge. The experimenter insisted that the difference between sludges should be discernible, so she asked to see the physical plots (the cement tubs).

In some of the tubs there were eucalyptus saplings, the roots of which would obviously have grown through the drainage holes in the bottoms of the tubs. In other tubs there were broad-leafed deep-rooted weeds. But in some tubs there were only relatively small plants.

A single deep-rooted plant would be expected to have an effect on the populations of microbes in a tub which was quite large compared to the difference between the types of sludge. In order to investigate the effect of types of sludge the number of deep-rooted plants should have been standardized, possibly by deliberate planting as well as having a standardized weeding regime.

5.5.2 Whether to Randomize

Randomization is needed if there may be correlated noise.

Randomization is often expensive – it may be cheaper to run experiments in a particular order. If this is the case, consider the costs and the possible ambiguity at the end. However you should always be able to give very good reasons if you do not randomize the order of performing runs in an experiment. The prima facie case for randomization is very strong. It tends to even out all of the possible biases that you haven't considered or haven't allowed for.

When you do an experiment, analyse the results, and conclude that certain factors have certain effects on some process, there is always some doubt about the correctness of your conclusions.

For instance, suppose you conducted a simple 3 by 2 factorial experiment consisting of the following six runs.

Run number	Factor A	Factor B
1	low	−
2	low	+
3	medium	−
4	medium	+
5	high	−
6	high	+

Suppose that the runs were conducted in the order indicated in the table. If the results came out to be in ascending order, then a formal analysis of the data would suggest that factors A and B both had a positive effect. However, the trend could easily be due to factors not included as such – things like ambient temperature, wear of equipment, or drift in the quality of raw materials.

Randomization of the order of doing runs greatly reduces the chance that factors not in the experiment will mess up the results. Sometimes it is also necessary that the order of testing be randomized. Ask yourself whether you are worried that a trend in the

PLAN A SINGLE EXPERIMENT

test results is likely and whether it might make you less confident of any interpretation of the results.

Sometimes randomization of the order of doing runs greatly increases the cost of doing an experiment. If changing levels of one factor is expensive then a split-plot experiment might be considered so that the factor needs to be changed less often than the other factors. However this will generally result in a reduction of the precision to which that factor can be measured.

The ultimate test as to whether randomization is necessary is to ask whether the failure to randomize would result in problems with selling the conclusions.

An important use of randomization is in double-blind experiments used for testing the efficacy of drugs in situations where there is likely to be a substantial placebo effect (tendency for people to get better because they believe that they have been given a useful drug). Neither patients nor doctors know which patients are being given the drug to be tested and which are given the placebo (sugar or coloured water).

Randomization may also be used in allocating batches of raw material to various treatments or trials. In some situations this may be more important than randomizing the order of performing the trials.

Example: Lanarkshire milk experiment

'Student' (1931) criticized an experiment in which a large number of school children in Lanarkshire were given one of three treatments:

- three quarters of a pint of raw milk daily for four months;
- three quarters of a pint of pasteurised milk daily for four months, or;
- no dietary supplement.

Between 200 and 400 children aged 5 to 12 years were chosen in each of 67 schools. Only one type of milk was used in each school. The intended experimental protocol included randomizing the assignment of children to the treatments using either an alphabetical system or a ballot. However, it became clear when the data were later analysed that some teachers had made an effort to ensure that the children more in need were given milk, although the official report only admitted 'In any particular school where there was any group to which these methods had given an undue proportion of well-fed or ill-nourished children, others were substituted in order to obtain a more level selection.' Clear evidence of bias is provided by the heights and weights of the children at the start of the experiment before they had received any dietary supplement. On average, the children assigned to receive milk were smaller in weight by three months development and smaller in height by four months development than the children assigned to receive no dietary supplement.

Such altruism on behalf of teachers is not surprising. Gore and Altman (1982, page 9) in discussing this experiment also highlight the fact that the children were weighed fully clothed. The weather was warmer at the end of the experiment than at the start so the average recorded weight gain over the four months of the experiment would be less than the average real weight gain, and it is likely that poorer children would have had less weight of clothes in winter so this bias is likely to be different between the children given milk and the children not given milk.

5.5.3 How to Randomize

A simple way to randomize the order of runs in an experiment is to write random numbers onto records which include all information about experimental runs to be performed and then to sort these records into order according to the random numbers. For a replicated experiment the replicates should be randomized separately.

Example

Three replicates of a 2^3 experiment are to be randomized. The levels for the three factors A, B and C are indicated by $-$ or $+$, for the lower and upper of the two possible levels. Random numbers are added to a table which gives the treatment combinations in a standard order. Randomization within each of the three replicates separately is achieved by sorting the rows of the table using replicate as primary sort key and the random number as secondary sort key. The result of such sorting is given in Table 1.

This randomization might result in some undesirable patterns. For instance, one treatment combination might be first in all replicates. It might be thought worth the effort of using the first three rows of a randomized 8×8 Latin Square to randomize the order.

5.5.4 Removing Sources of Variation by Statistical Analysis

Sometimes sources of variation can be removed by looking at changes rather than at absolute values. The example below on rats losing weight (see page 139) illustrates this. In many circumstances the effects of factors which can be measured but not controlled can be removed by statistical analysis. The factors which can be measured but not controlled are referred to as covariates.

It may be important to plan to measure covariates. For instance, suppose that you intend to conduct some experiments next year comparing plant varieties and fertilizer treatments. You could plant a standard crop this year on all of the plots which will be used next year, and measure the yield of that crop on each plot. These measured yields will provide a measure of the fertility of the various plots and could be used as a covariate in the analysis of next year's experiments.

Example: Gold recovery

Meyer and Napier-Munn (1999) described an experiment conducted at Bougainville Copper Limited in 1986. The purpose of the experiment was to see whether a new flotation reagent would improve gold recovery without compromising copper recovery. Over a period of 112 days the two alternative flotation reagents were used on alternate days.

We will consider only the data on gold recovery, the percentage of gold in the ore which is recovered by the metallurgical processes. One way to analyse the data is to use a paired t-test. Data is also available on the feed grades. This data ought to be used as a covariate since it is expected that recovery will be higher when the feed grade is higher.

PLAN A SINGLE EXPERIMENT

Table 1. Example of an experiment in which the order of experimental runs has been randomized within replicates

Standard sequence number	Replicate	Random number	Level of factor A	Level of factor B	Level of factor C
3	1	0.2002	+	+	−
6	1	0.3317	−	−	+
8	1	0.3983	−	+	−
1	1	0.4967	+	+	+
4	1	0.5538	+	−	+
5	1	0.6477	+	−	−
7	1	0.8922	−	−	−
2	1	0.9903	−	+	+
9	2	0.0869	−	+	+
10	2	0.4526	+	−	−
15	2	0.5225	+	+	+
13	2	0.6469	−	−	+
16	2	0.8665	+	+	−
11	2	0.8796	−	+	−
12	2	0.9135	+	−	+
14	2	0.9418	−	−	−
21	3	0.1194	+	+	+
18	3	0.2550	+	−	−
24	3	0.4587	−	+	+
23	3	0.4934	+	+	−
22	3	0.6519	−	−	−
20	3	0.6630	−	+	−
17	3	0.8308	+	−	+
19	3	0.8450	−	−	+

Figure 4 shows one way to look at the data for the 56 pairs of experimental runs. There is a tendency for the recovery using the new reagent to be higher when the feed grade is higher for that day's run. The regression line shown as a dashed line on Figure 4 has a slope of 0.040 with a standard error of 0.010 and an intercept corresponding to equal feed grades of 1.41 with a standard error of 0.60. The intercept is the difference between the reagents which is of interest.

Figure 5 shows that there is another source of noise in this data which might be considered when planning future experiments. On this graph the triangles tend to be higher up than the crosses, suggesting that better extraction for a given feed grade was achieved early in the experiment than late in the experiment. There is probably some other property of the ore which varied slowly over the period of the experiment and which could be used as another covariate.

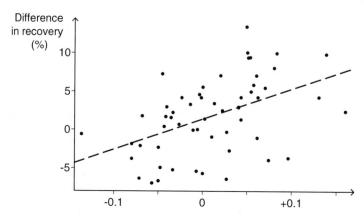

Figure 4. Scatter plot of differences in recovery against differences in feed grades for 56 pairs of runs. Differences are calculated as data relevant to the new reagent minus data relevant to normal reagent

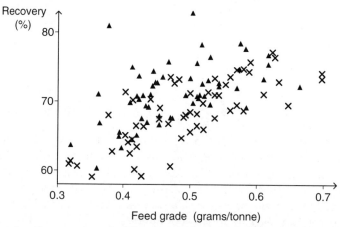

Figure 5. Scatter plot of recovery against feed grades. Data for the first 56 days of the experiment are plotted as solid triangles. Data for the last 56 days of the experiment are plotted as crosses. Note that triangles tend to be higher on this graph than crosses

PLAN A SINGLE EXPERIMENT

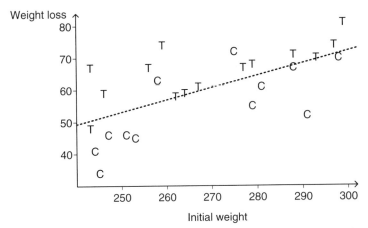

Figure 6. Weight loss for treatment (T) and control (C) rats plotted against initial weight. The dashed line shows the a linear regression of weight loss against initial weight. The points for treated rats are generally above this line and those for control rats are generally below

Example: Rats losing weight

This example was passed on to me by a colleague, Albert Trajstman.

In a situation which would normally cause rats to lose weight, some rats were treated in a way which was thought likely to reduce the weight loss over a week. These rats had initial weights (in grams) of 288, 243, 267, 277, 256, 264, 243, 299, 293, 297, 279, 259, 246 and 262. Their final weights were 217, 195, 206, 210, 189, 205, 176, 218, 223, 223, 211, 185, 187 and 204. Other rats which were not treated had initial weights of 279, 245, 291, 281, 253, 244, 298, 258, 251, 247, 288 and 275. Their final weights were 224, 211, 239, 220, 208, 203, 228, 195, 205, 201, 221 and 203.

A first, rather naïve, analysis of this data is to do Student's t tests on both initial and final weights. The t test on initial weights shows no statistically significant difference ($t = 0.26$). This could be regarded as evidence that the experimenter allocated the rats to treatment and control groups at random. The t test on final weights also fails to show a statistically significant difference ($t = 1.74$).

However, some of the variation in the final weights is related to the variation in initial weights. This suggests that we might look at weight losses. The average weight loss is 54.3 for the treated rats and 66 for the for the untreated (control) rats. A t test on weight losses does show a statistically significant difference ($t = 2.86$).

A more sensitive analysis uses a model which says that the weight loss varies linearly with the initial weight and is also affected by the treatment. The t statistic for the statistical significance of the treatment effect is 3.80. This is statistically significant at the 0.1% level.

Figure 6 shows that the data for this example, together with a line showing the estimated effect of initial weight on weight loss averaged over treatment and control rats. Heavier rats tend to have a larger weight loss. The statistical significance of the effect of the treatment can be thought of as summarizing the tendency for the Ts on this graph tend to be high relative to the line while the Cs tend to be low relative to the line. This tendency would be difficult to deny.

5.6 CASE STUDY: SURFACE TREATMENT OF TOOL STEEL

There are a wide variety of processes used in metalworking and manufacturing industry which are intended to improve the life of metal tools and metal parts, including gears, die casting tools and hot forging tools. These processes can be broadly classified as surface coatings and diffusion treatments. The surface layers produced have a variety of chemical compositions, including chromium, nickel, boron, silicon and carbides, nitrides and carbonitrides of iron, titanium, vanadium and chromium.

Desirable properties of the surface layers include wear resistance, indentation resistance, fatigue strength, corrosion resistance and surface finish. Three weaknesses of some of the processes for certain purposes are as follows.

- Processes which involve high temperatures often distort the metal parts which are treated or coated.
- Some surface coatings are not sufficiently well bonded to the metal substrate for some applications.
- Processes which deposit material by moving atoms in straight lines from a source are difficult to use with parts of complex geometry.

A company called Quality Heat Technology has a new process (Qab) which uses a fluidized bed to diffuse a nitrocarburized layer onto the steel substrate, simultaneously nitrocarburizing the outermost layers of the substrate. The bed being used experimentally is cylindrical in shape. It is about 300 mm diameter and about 600 mm high. It consists of particles about 100 μm in diameter of alumina and chromium. The alumina which is 90% by weight is thought to be inert.

Fluidizing gases are introduced at a total flow rate of about 5 m^3 per hour and are typically composed of about:

- 1–3% HCl (commonly 1.5%);
- 10–35% NH_3 (which is normally about 200 ppm H_2O but can be up to 500 ppm H_2O);
- CO_2 (sometimes 1% liquefied petroleum gas which is propane or C_3H_8 is also used);
- enough N_2 for the total flow rate required to keep the bed fluidised.

The gases are introduced into the base of the fluidized bed by special diffusers protected by larger particles of Al_2O_3 to allow the free flow of gases but to avoid combination of some gases until they have been heated and to protect the diffusers from heat. The bed is heated to a prescribed temperature, say 570° C, the work is immersed in the bed and the gases flow for about 5 hours. Then the workload is removed and cooled.

The Qab process diffuses carbon, nitrogen and chromium into the outermost layers of the substrate steel, to form a 'nitrocarburized layer'. It also adds a thin coating, referred to as the 'white layer', which is composed mainly of chromium carbonitride and chromium carbide, as well as a conventional iron–carbon–nitrogen nitrocarburizing layer.

PLAN A SINGLE EXPERIMENT

5.6.1 *Objective*

Scope

Experimentation adequate to demonstrate the feasibility of the process has been done. The process seems to work on a variety of steels including stainless steels. However, hardnesses achieved have seldom exceeded 1400–1500 VHN and the people involved had hoped to achieve 1700–1800 VHN.

A marketing study has been done which suggested that the world-wide market for the process would be worth about $200 million per annum. Much of this market is for treating tools, so it seems reasonable to concentrate on tool steels in the immediate short term (say, six months).

Quality Heat Technologies has two provisional patents. The issue of most immediate interest is to convince the board of a joint venture to spend about $200 000 to take out world-wide patents.

Another issue is to differentiate the process from patents currently in place by Toyota and others. This requires showing the importance of controlling nitriding potential, chromium potential, carbon and oxygen potential.

It is desirable to check whether the process can be used with vanadium, tungsten, titanium and other like metals.

5.6.2 *Beliefs and Uncertainties*

A summary of beliefs and uncertainties was drawn up by three of the staff of Quality Heat Technologies.

The focus should be on the metal transfer into the white layer (e.g. Cr or V). This is the novelty of the process and sets it apart from other processes.

Theory

HCl activates the metal surface. It also reacts with NH_3 to form ammonium chloride. Using too much HCl produces a porous, relatively thick layer of a red coating on the surface of parts made from low alloy or plain carbon steels.

At low temperatures nitrogen diffuses faster than carbon, which diffuses faster than chromium. The driving force for chromium diffusion is not definitely known. The mechanism of transfer of chromium is thought to involve $CrCl_3$. Perhaps a similar effect could be obtained using $CrCl_3$ salt.

Greater time would be required for diffusion at lower temperatures. Using high temperatures reduces the strength of substrate steel. The process operating temperature should not be greater than the normal tempering temperature for the steel. Time affects process profitability, so it is best to use the highest temperature which does not reduce the strength of the substrate or cause unacceptable levels of distortion.

Experimental results to date

When I saw a list of the experimental results, 86 experimental runs of the fluidized bed had been performed. Early trials in which the surface of the substrate was nitrocarburized first and chromium was diffused into the surface later were not successful in producing useful surface coatings. The detailed results are not important to the points that I wish to make using this example, so they are not described here.

Measurement

The issue which is the greatest barrier to progressing this development is the difficulty of measuring what has been achieved.

The method of measurement which has been most useful is a microhardness test which involves making an indentation with a diamond. This has the advantage over some other measurement methods that it can be used to test the hardness at various depths from the surface. Thus the white layer and the nitrocurburised layer can be separately tested.

Testimonials of customers about the performance of tools treated using the Qab process have been extremely encouraging, but they are not able to differentiate slight differences between variations of the process and so help in its refinement.

Measurements have also been made on the coefficient of friction when Qab treated tools are used on mild steel, wear resistance, corrosion resistance, seizure resistance, peel strength or adhesivity, impact resistance, the thickness of the white layer and the thickness of the diffusion layer.

The measurement method which seems most capable of distinguishing between slight variations of the process is referred to as GDOS. This measures the local chemical composition of white layer over a very small region. Layers of around a micron in thickness can be removed by gas discharge and the chemical analysis repeated. Scanning electron microscope techniques have not been found to be as useful.

5.6.3 Strategy

The proposed strategy is to first look in what is currently regarded as the most promising region to see if improvement is possible. Crucial elements of strategy are to concentrate on tool steels first and to only look at chromium in the short term. We intend to look at temperatures near 570 °C first, since this is satisfactory for tool steel. We expect to investigate other temperatures at a later time.

Reduction in cost is not important compared to the speed of getting to the market. The most important strategic decision has been to try to get access to a local GDOS machine. The nearest available GDOS machine is in the USA. Getting samples tested takes a substantial and variable time. It is also very difficult to check that operators of the machine use appropriate standards and that they understand the nature of Quality Heat Technology's quest.

Experimentation is currently proceeding slowly, but should proceed quickly once a local GDOS machine becomes available in about July 2000.

The main point of this case study is that measuring can be critical.

6

Design the Experiment

This chapter deals with the selection of factors, the selection of levels for those factors and the selection of a set of trials to be conducted. It starts with some relatively theoretical material about the concepts of confounding and orthogonality, how to estimate how large an experiment will need to be and the types of experimental designs which are available.

6.1 THE CONCEPTS OF CONFOUNDING AND ORTHOGONALITY

Two factors (or, more generally, two possible ways of explaining something) are said to be completely confounded for a given set of data if it is not possible to distinguish between them using that data, because a model with one factor and a model with the other factor give exactly the same predicted values for all data points. This means that trends in the data which may be observed cannot be reliably attributed to one factor being investigated.

Orthogonality is the opposite extreme from confounding. Two factors are said to be orthogonal for a set of data if the estimation of each of them is unaffected by the other factor.

The concepts of confounding and orthogonality are basic to the theory of designing experiments. In my experience, attempts to give precise mathematical definitions of them is not helpful. This section attempts to explain these concepts by using examples.

An example of complete confounding: washing powder advertisements

Imagine a television commercial showing the virtues of brand A washing powder relative to brand X: Mrs. A's childrens' clothes look bright and sparkling after being washed using brand A. Next door, Mrs. X's childrens' clothes look rather drab after being washed using brand X.

Unless our critical faculties have been numbed by the various methods that the advertising and television industries are constantly refining, we recognize that the difference in the appearance on the clothes may be due to factors other than the difference in the brand of washing powder.

- Mrs. A's children are both girls who enjoy indoor pursuits. Their clothes are mostly brightly coloured and made from artificial fabrics. Mrs. A has an almost new

ultrasonic-assisted type AAA washing machine and always uses water softener in the second rinse.
- In contrast, Mrs. X's children are three boys who play a lot of very rough outdoor games. Mrs. X does not use water softener or buy clothes made from artificial fabrics. She encourages her children to choose clothes of colours that don't show the dirt. Her washing machine is a conventional one, and rather old.

To restate this as an example to illustrate the meaning of 'confounding': if the only information available to us is the average cleanliness of Mrs. A's and Mrs. X's washing, then the effect of washing powder is completely confounded with, for instance, the effect of using water softener.

6.1.1 Confounding and Orthogonality of Estimates of Trends

Usually, when people talk about confounding and orthogonality they are referring to trying to estimate parameters in models for trends. Let us consider the simplest situation for which confounding and orthogonality are possible: data which may be influenced by two factors, which we will call factor A and factor B.

Figure 1(a) illustrates a completely confounded experimental design. Given response data at the two design points, it would be impossible to tell whether factor A or factor B is having an effect, or whether they both have effects. Intuitively, this design is like that used in the washing powder example above. Formally, many people would not use the term 'experimental design' to refer to such a plan, because it is obvious that the data obtained will not tell a straightforward story in the way that we expect experimental data to do. Computationally, attempts to analyse such data, for instance to fit a linear regression model including terms for both factors, can be expected to generate some sort of error message.

Figure 1(b) illustrates a perfectly orthogonal experimental design. The levels of factor A are used equally often (in this case, once) for each level of factor B. Also, the levels of factor B are used equally often for each level of factor A. If we take some response data at the four design points and fit a linear regression model then we will find that the correlation between the estimated effects of the two factors is exactly zero. This means that there is no confounding between the estimates of the two effects. The best estimate of the effect of factor A does not depend on the estimate of the effect of factor B and would not be changed if factor B were omitted from the model. Similarly, the best estimate of the effect of factor B does not depend on the estimate of the effect of factor A and would not be changed if factor A were omitted from the model. (The regression model being fitted is assumed to have an intercept. If it does not then the estimated effects will negatively correlated. It might also be noted that the estimated effects are correlated with the estimated intercept. This correlation is not important from a conceptual point of view. It can be removed by changing the origin of the scales of the factors so that the average level of each of the factors is zero.)

Figure 1(c) illustrates an almost perfectly confounded experimental design. Given response data at the six design points it is almost impossible to tell whether factor A or factor B is having an effect, or whether they both have effects. A linear regression model reported a correlation between the estimated effects for the two factors of -0.98.

DESIGN THE EXPERIMENT

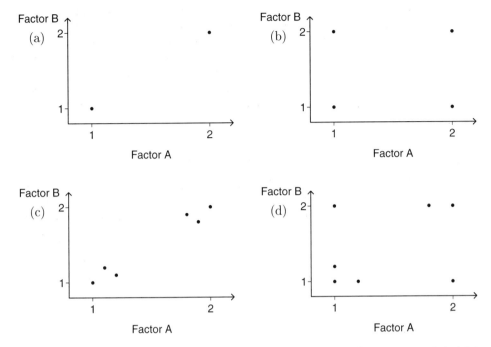

Figure 1. Four possible designs for an experiment in which two factors are varied: (a) is completely confounded, (b) is orthogonal, (c) is almost confounded and (d) is unbalanced but near enough to orthogonal for practical purposes

Figure 1 (d) illustrates an experimental design which is near enough to orthogonal for most practical purposes. A linear regression model reported a correlation between the estimated effects for the two factors of -0.29. It is negative because if the effect of factor A seems larger then the effect of factor B seems smaller.

Some practical situations

Some examples of situations where thinking about confounding and orthogonality might be helpful are as follows.

Selecting cows from a dairy herd for comparing antibiotics for treating mastitis could be done by treating the first half to arrive at the milking shed with one antibiotic and the second half with the alternative antibiotic. What would be wrong with this? The order of arrival of cows at the milking shed might be related to general health (sick cows being slow, or sick cows tending to hang about the shed) to age and to social status within the herd. These possible effects should not be allowed to be confounded with the comparison of interest.

Suppose two options on a machine for making widgets were compared by using one option every morning shift and the other every afternoon shift. The difference between the options would be confounded with the difference between shifts.

Suppose that patients volunteering for an experiment have an opportunity to withdraw after they find out which of two treatments they will receive. Unless the number of such withdrawals is very small, any difference in average efficacy of the treatments will not be trusted because it is confounded with possible differences in average prognosis between people inclined to withdraw and people not inclined to withdraw. A better procedure is to ask well-informed patients to sign a form saying that they would be willing to receive either experimental treatment and to select patients from those that have signed.

6.1.2 Confounding between Main Effects and Interactions

The type of confounding that is discussed most in books about the design of experiments is confounding between main effects and interactions or between two different interactions. Two effects or interactions are said to be confounded for a given set of data if it is not possible to distinguish between them using that data because they give the same predicted values.

Here is a simple example to illustrate this idea, for the simplest case of a main effect being confounded with a two-factor interaction. The general idea will be discussed further when we discuss fractional factorial experimental designs on page 156.

Example: Aircraft safety and logos

An example of complete confounding between a main effect and an interaction is an experiment on aircraft design with the following three factors.

A: Length of the left-side wing of an aircraft. Levels for the factor are 25 and 30 metres.
B: Length of the right-side wing of an aircraft. Levels for the factor are 25 and 30 metres.
C: Whether the 'Fly-By-Night No Liability' airline logo has been painted onto the fuselage. Levels for this factor are 'Yes' and 'No'.

Four experimental aeroplanes are built according to a fractional factorial design.

Length of left wing (m)	Length of right wing (m)	Is logo present?
25	25	Yes
25	30	No
30	25	No
30	30	Yes

The first and last of these experimental aeroplanes were found to fly well. The second and third were both rated 'extremely unstable' by test pilots and crashed during testing by ordinary pilots. One obvious way of explaining these outcomes is that the 'Fly-By-Night No Liability' logo is crucial. However, there is another possible, more complicated explanation. There may be an interaction between the first two factors. One way to explain such an interaction is to say that the effect of an aeroplane having a 25 m long left wing rather than a 30 m long left wing depends on the length of the right wing. (In other words, wings of the same length are better than wings of different lengths.)

DESIGN THE EXPERIMENT

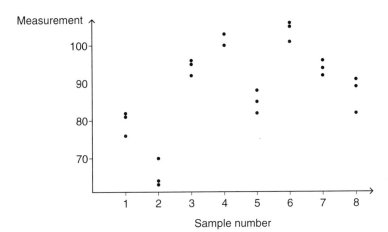

Figure 2. Plot of replicate measurements on eight samples

6.1.3 A More General Caution

The idea behind the concepts of confounding and orthogonality is that when fitting models to data you need to be cautious about whether the various features of models can be reliably differentiated.

For instance, consider an experiment in which eight replicate samples were taken and a single measurement was done on each sample. The variation between the results could be entirely due to sampling variation or could be entirely due to testing variation. The two components of variation cannot be reliably differentiated.

In contrast, suppose that the measurement procedure was replicated three times on each of the eight replicate samples, giving data like the following:

$$\begin{array}{cccccccc} 76 & 64 & 92 & 103 & 85 & 101 & 92 & 89 \\ 82 & 70 & 95 & 100 & 88 & 106 & 94 & 91 \\ 81 & 63 & 96 & 103 & 82 & 105 & 96 & 82. \end{array}$$

The average of the sample variances of the sets of three replicate measurements is

$$(10.3 + 14.3 + 4.3 + 3.0 + 9.0 + 7.0 + 4.0 + 22.3)/8 = 9.3.$$

This is quite a good estimate of the measurement variance.

The average measurements on the eight samples are 79.7, 65.7, 94.3, 102.0, 85.0, 104.0, 94.0 and 87.3. The variance of these numbers is 156.8, which is a good estimate of the sampling variance plus $\frac{1}{3}$ of the measurement variance. Hence a good estimate of the sampling variance is 153.7.

Figure 2 shows these results. It is easy to see that the sampling variation can be separated from the measurement variation.

Consider the data shown in Figure 3 which is from Draper and Guttman (1980). The data are yields of barley in tonnes per hectare for a sequence of 38 plots. The factor being investigated was the spray frequency of tridemorph. Its levels were denoted by '0', '1', '2' and 'R', with '0' being no application. That yields were generally lower when tridemorph was not applied is easy to see from Figure 3.

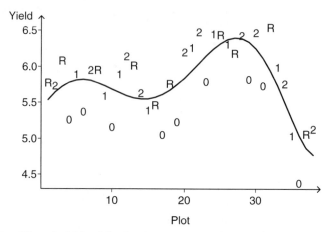

Figure 3. Plot of yields of barley in tonnes per hectare for Rothamstead mildew control trial 75/R/B/9. The trend line shown is a polynomial of order six. The plot symbol indicates the treatment

This particular data set has been analysed in a large number of ways by various people. They have fitted models with features like the following.

- Using the average yield (adjusted for their treatment means) of adjacent plots as a covariate.
- Fitting fertility trends, like the polynomial shown in Figure 3 but modelled in a variety of ways. One of these ways is to use splines.
- Modelling variation in fertility between plots as Brownian motion or as a time series model with a structure and correlations to be estimated from the data.
- Fitting models in which the yield is expected to be affected by the treatments applied to neighbouring plots.

I do not wish to express an opinion about the values of the various approaches. However, I suggest that great caution should be exercised when fitting models which combine more than one of these features, because I doubt that the various features of such complex models could be reliably differentiated.

6.2 ISSUES AFFECTING THE NUMBER OF RUNS

There is no single, simple answer to the question 'How many runs do I need?' In practice, choices of experimental design and numbers of runs are frequently made on the basis of experience with similar experiments. There is nothing wrong with this. For some experiments, more detailed consideration of the number of replications is necessary.

The choice of the number of runs in an experiment will generally be influenced by strategic issues such as the costs and benefits of experimenting and whether there are substantial uncertainties about the experimental technique, components of variability and the region of interest, as was discussed on page 61.

DESIGN THE EXPERIMENT

This section assumes that those issues have been confronted. It gives three technical answers to the question of how many runs should be performed and provides some guidelines at to when each of those answers is likely to be useful.

6.2.1 First Answer: Enough Runs to Estimate the Components of Variability

The first answer to the question 'How many runs?' is applicable to experiments investigating sampling and testing errors. Such experiments are designed to estimate components of variance. They need to be large enough for those components of variance to be estimated to enough precision.

Figure 4 indicates how the precision with which a component of variation is likely to be estimated varies with the amount of information used to estimate that component.

For instance, suppose that you planned to conduct an experiment in which an aliquot of material was divided into eight samples which were then tested. The eight data points would provide seven degrees of freedom for estimating the sampling and testing error variance. From Figure 4 you can see how accurate your estimate, s, of the true sampling and testing error standard deviation, σ, is likely to be. The 10% and 90% quantiles of the expected distribution of s/σ are about 0.64 and 1.31. So there is an 80% chance that s/σ will be between 0.64 and 1.31. If the aliquot were divided into sixteen samples then there would be 15 degrees of freedom, so the 10% and 90% quantiles of the expected distribution of s/σ would be about 0.75 and 1.22.

These results can be calculated from the readily available quantiles of the χ^2 distribution, using the relationship that when there are f degrees of freedom fs^2/σ^2 has a χ^2 distribution with f degrees of freedom. I have plotted the quantiles of s/σ in Figure 4 because I believe that it is more immediately comprehensible than the χ^2 distribution. Hald (1952) tabulated the quantiles of s^2/σ^2, but this is also less immediately comprehensible in my opinion.

It should be noted that the number of degrees of freedom for estimating a component of variance may be quite different from the number of runs in an experiment. For instance, in an interlaboratory trial involving n laboratories there are only $n-1$ degrees of freedom for estimating the between-laboratory component of variation, no matter how many results are measured within each of the laboratories. For components of variation which can be estimated using within-laboratory information there is generally plenty of information.

Some qualifications to the use of this approach are as follows.

- Is it likely that some experimental runs will be lost? If so, a few extra runs beyond the number otherwise required will provide some insurance.
- Do you want to be able to check for outliers or to estimate the amount of variability in a robust way which is not much affected by such outliers? If so, you need to be even more conservative.
- When using data with a hierarchical structure, such as that from interlaboratory trials, rejection or loss of data may occur at the highest level. For instance, one laboratory may decline to participate or may produce results which are well outside the pattern of results produced by other laboratories.

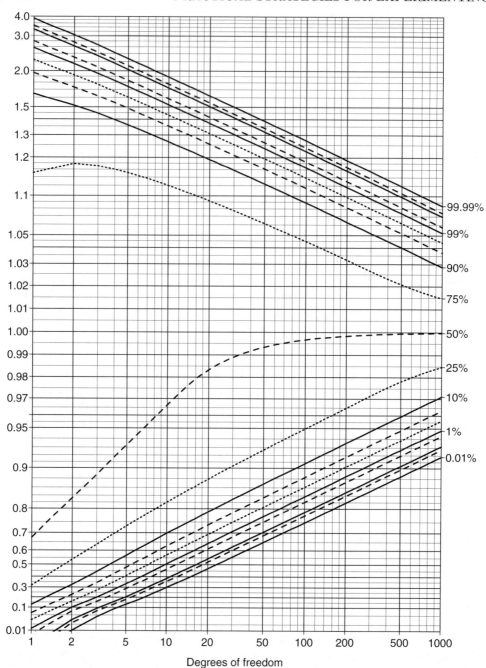

Figure 4. Quantiles of the ratio s/σ where s is the sample standard deviation for a sample from a normal distribution with population standard deviation σ. Values for s/σ are labelled on the left and some quantiles are labelled on the right. The quantiles drawn are those for probabilities 0.01%, 0.05% (dashed), 0.1%, 0.5% (dashed), 1%, 2.5% (dotted), 5% (dashed), 10%, 25% (dotted) and 50% (dashed), plus those for the complementary probabilities

DESIGN THE EXPERIMENT

6.2.2 Second Answer: Enough Runs to Reliably Make a Simple Choice

You might anticipate making a simple choice between two levels of a factor. Suppose that you would like to be able to make a reliable choice even if the true difference in mean response for those levels of the factor was only some minimum amount of concern, say, δ. A useful, approximate formula for the number of replications (n) which are needed for the levels of the factor is

$$n = 2[(Z_{\alpha/2} + Z_\beta)\sigma/\delta]^2$$

where $Z_{\alpha/2}$ is the two-tailed critical value from a standard normal distribution corresponding to the intended statistical significance level α, Z_β is the one-tailed critical value from a standard normal distribution corresponding to a pre-selected probability β of failing to detect a difference δ, and σ is the standard deviation associated with single measurements.

This formula can be derived by noting that a difference between single observations at the two levels of the factor will have variance $2\sigma^2$. Hence the average of n replicate differences will have variance $2\sigma^2/n$ and standard deviation $\sqrt{2/n}\,\sigma$. If the true difference is δ, the observed difference will exceed $\delta - Z_\beta\sqrt{2/n}\,\sigma$ with probability $1 - \beta$, and this needs to be at least $Z_{\alpha/2}\sqrt{2/n}\,\sigma$ in order to be statistically significant.

This formula neglects the uncertainty in the estimation of error variance, so if it gives you an n less than 10, try to use an extra one or two replications.

One common choice is 0.05 for α, corresponding to the use of the 95% statistical significance level, and 0.1 for β, so that there is a 90% probability of picking up a difference as small as δ. This gives $Z_{\alpha/2} = 1.96$ and $Z_\beta = 1.28$ and hence

$$n = 21(\sigma/\delta)^2.$$

This means that if you want a 90% chance of getting an actual difference of δ between two treatments to be detected with statistical significance at the 95% level, then you need to use at least $21(\sigma/\delta)^2$ replications when the error standard deviation is σ.

Another possible choice is 0.1 for α, corresponding to use of the 90% statistical significance level, and 0.5 for β. This gives $Z_{\alpha/2} = 1.645$ and $Z_\beta = 0.0$ and hence

$$n = 5.4(\sigma/\delta)^2.$$

This means that by using $5.4(\sigma/\delta)^2$ replications a difference of δ between two treatments will be statistically significant at the 90% level. An actual difference of δ will lead to only a 50% chance of having an observed difference of δ.

When endeavouring to apply these formulae, a value for σ will often be available from the results of previous experiments or from analysis of historical data. If it is very difficult to make any sort of guess about the value of σ, then a preliminary experiment should be undertaken.

Choosing a value for δ to put into this formula often seems difficult. In many circumstances, δ could be taken to be the smallest difference thought economically important enough to justify the taking of some action.

6.2.3 Third Answer: Enough Runs to Optimize a Continuously-Variable Factor

A philosophically sensible general method for answering the question 'How many runs?' is to use a Bayesian method called pre-posterior analysis. This involves expressing current opinions and uncertainties about unknown parameters by using prior probability distributions, considering the data that might occur if an experiment were conducted and evaluating the average utility of the states of knowledge which might occur. Regrettably, the method tends to require a lot of complicated calculations. The effort of performing such calculations doesn't seem justified when the information fed into the method consists only of guesses.

A simple case of such calculations runs as follows. Suppose that we make $3n$ observations in an experiment in which a response Y is measured. There is only one factor, X, with three levels, $x_0 - h$, x_0 and $x_0 + h$. The mean response is a smooth function of X which is approximately equal to the quadratic function $A - R(x - \theta)^2$ near the optimum value θ. The standard deviation of the response is σ. The marginal cost of observations is u, and the loss incurred by running the process with X at the best estimate $\hat{\theta}$, rather than at precisely θ, is $C(\hat{\theta} - \theta)^2$.

The difference of average values $\bar{Y}_{x_0+h} - \bar{Y}_{x_0-h}$ has variance $2\sigma^2/n$. An estimate of the gradient of the mean response, $-2R(x - \theta)$, at x_0 is the slope of the line joining $(x_0 - h, \bar{Y}_{x_0-h})$ and $(x_0 + h, \bar{Y}_{x_0+h})$, so an estimate of θ is $x_0 + (\bar{Y}_{x_0+h} - \bar{Y}_{x_0-h})/(4hR)$. This has variance $(2\sigma^2/n) \div (4hR)^2 = \sigma^2/(8nh^2R^2)$. Therefore, the expected loss is $C\sigma^2/(8nh^2R^2)$ and the total expected cost is $3un + C\sigma^2/(8nh^2R^2)$. Setting to zero the derivative of this with respect to n tells us that the value for n which minimizes expected loss is

$$n = \sqrt{\frac{C}{24u} \frac{\sigma}{hR}}.$$

As an illustration of the application of this formula, suppose that a metallurgical process appears from some preliminary trials to run at a yield of about 86% at a temperature of 210°C but the yield is only about 84% at 200°C and 220°C. Trials cost $100 each and have a standard deviation of about 2% on the yield scale. The quadratic term in the response function is approximately $0.02\%(°C)^{-2}$. If the net present value of 1% extra yield is said to be $15 million then, since a 1°C error causes a yield loss of 0.02%, $C = \$300\,000\,(°C)^{-2}$. If we experiment with the temperatures 200°C, 210°C and 220°C then $h = 10°C$. Using the formula above, we calculate

$$n = \sqrt{\frac{C}{24u} \frac{\sigma}{hR}} = \sqrt{\frac{300\,000}{24 \times 100} \frac{2}{10 \times 0.02}} = 112.$$

Having done such a calculation we ought to check its reasonableness. Two points seem to be particularly relevant here.

The large value used for C comes from the large monetary value of differences in yield. The $15 million figure was calculated using the entire economic life of the plant. It would be more realistic to assume that some full-scale trials will be conducted to fine-tune the operations of the plant, with the planned experiment only affecting the output of the plant for about a month. A more realistic value for C is $\$3000\,(°C)^{-2}$, which would change n to 11 or 12.

The most crucial assumption is that the response curve is quadratic over the range being investigated. This might be unreasonable over a wide range, but is likely to

DESIGN THE EXPERIMENT

be reasonably accurate over a small range. If we had planned to use a narrower experimental range, say from 205°C to 215°C, then the experiment would need to be proportionately larger.

Remark: When experimenting by making slight modifications to normal operations, the cost of experimenting is often not primarily determined by the number of trials but is greatly affected by the likely loss in production or the deterioration in quality compared to the best known operating conditions. This means that experiments with $h = 5$ would be only about a quarter as costly as experiments with $h = 10$. Twice as many trials would be required in order to estimate the best operating conditions with the same precision. (Actually, more than twice as many because the assumption that the second derivative of the response surface is known is actually not true, and determining that second derivative to the same precision requires four times as many trials.) This illustrates the general principle that small changes from standard conditions should be used when experimenting at production scale.

6.2.4 First Answer Revisited: Is the Error Variance Known Well Enough?

For experiments in which the effects of many factors and interactions are being estimated, it is generally desirable that there be sufficient information about the error variance in order to estimate it reasonably well. In some experiments within a sequence of experiments it may be adequate to assume that the error variance is the same as in earlier experiments, but even in such circumstances the monitoring of estimates of error variance is useful for checking that experimental precision has not deteriorated.

The precision with which the error variance can be estimated from an experiment in which a complicated model for the trend is also estimated can be considered by looking at the form of the analysis of the variance table which will be produced after the experiment has been run. For experiments in which the factors have discrete, unordered levels (like varieties of wheat, types of machine, suppliers or operators) the following steps are likely to be useful.

- The total number of degrees of freedom will be the number of observations made less one degree of freedom for the overall mean.
- The main effect for a two-level factor uses one degree of freedom.
- The main effect for a three-level factor uses two degrees of freedom.
- The main effect for an m-level factor uses $m - 1$ degrees of freedom.
- Blocked experiments can be regarded as having an additional factor with the number of levels equal to the number of blocks, so blocking uses one less degree of freedom than the number of blocks.
- An interaction between two two-level factors uses one degree of freedom.
- An interaction between an m-level factor and an n level factor uses $(m-1)(n-1)$ degrees of freedom.
- If individual two-factor interactions are not to be estimated but it is desired that they not be confounded with main effects (i.e. the experiment has resolution IV) then the number of degrees of freedom associated with interactions can be as small as the number of degrees of freedom associated with main effects.

For experiments with factors which have continuously variable levels on ordered scales (like temperatures, quantities and times), the number of degrees of freedom used by the model for the trend is equal to the number of parameters in that model. For a quadratic surface in p factors, there will be a constant, p linear terms, p quadratic terms and $p(p-1)/2$ cross-product terms, which is a total of $(p+1)(p+2)/2$ degrees of freedom. The table below indicates the number of parameters to be fitted for up to ten factors.

No. of factors	2	3	4	5	6	7	8	9	10
No. of parameters	6	10	15	21	28	36	45	55	66

The number of degrees of freedom available for estimating the error variance is the total number of data points less the number of parameters used to describe the trend.

- If this number is negative then it will not be possible to fit a model of the planned complexity to describe the trend.
- If this number is smaller than, say, 10, then the precision with which the error variance will be estimated will be rather poor. This may be quite acceptable if information about the error variance is available from other sources or if the purpose of the experiment is exploratory.
- If this number is greater than, say, 10, then the estimated error variance will be useful for judging the statistical significance of the conclusions reached. Such judgements compare things of interest with the estimated standard deviation, s. They would be compared to the true standard deviation, σ, if this were known, and Figure 4 tells us how different these alternative comparisons might be.

Mead (1988, page 587) calls the calculation of the number of degrees of freedom which will be available for estimating the error variance 'the resource equation'.

6.2.5 Remarks

For sequential experimentation, unless benefits are greatly affected by a time delay it is best to be cautious about doing large experiments.

The second answer to 'How many runs do I need?' tries to achieve statistical significance. This may be needed for publication, but not for selecting the best operating conditions.

Sometimes there are substantial practical constraints on the size of an experiment. This means that the cost of doing the experiment is not proportional to the size of the experiment. For instance, if a certain number of runs or trials can be done in a single day at a site remote from the experimenter's home base, then the number of trials might as well be a multiple of the number of trials that can be comfortably completed in a single day.

Sometimes the best sample size is zero – the value of the information likely to be obtained is less than the cost of obtaining it. This would be an important conclusion.

6.2.6 An Issue with Surprising Little Effect on the Size of Experiments – the Number of Factors

When an experiment has more than one treatment factor, the precision for estimating main effects is approximately the same as if the same amount of resources had

Table 1. Design for factorial experiment with two factors at three levels

Abstract levels		Practical application	
Factor A	Factor B	Temperature (°C)	Amount of X (g)
1	1	200	65
1	2	200	75
1	3	200	85
2	1	210	65
2	2	210	75
2	3	210	85
3	1	220	65
3	2	220	75
3	3	220	85

been devoted to a single-factor experiment. If all factors have two levels then the precision is undiminished. In practice, the error standard deviation is likely to be slightly larger because applying different levels of a treatment is more difficult than leaving a treatment unchanged. For factors with more than two levels, the precision for estimating interactions is less than for main effects.

If an experiment will be analysed as a split-plot design, then the precision for estimating effects at the plot stratum depends on the plot stratum error variance and the number of plots, but is little affected by the number of sub-plots per plot. Effects estimated in the sub-plot stratum are estimated as precisely as would be expected without worrying about the split-plot structure. Their precision depends on the sub-plot stratum error variance and the number of subplots. My advice is to seek professional help for designing experiments where there is more than one component of error variance.

6.3 SOME TYPES OF EXPERIMENTAL DESIGNS

6.3.1 Factorial Experiments

This section discusses factorial designs. These are experiments using all combinations of the chosen levels of the chosen factors. They are very practical and are easy to understand. Their major drawback is that they require too many runs when the number of factors is large (say, more than 3).

'Two-way layout' experiments is an alternative name for the simplest form of factorial experiment, involving two factors only. Each level of the first factor is used in combination with each level of the second factor.

An example of a factorial design is a 3×3 design which might be used as in Table 1. Sometimes the formal levels of factors are denoted by $1, 2, \ldots, n$. Sometimes the formal levels of factors with two levels are denoted by -1 and $+1$ and the formal levels of factors with three levels are denoted by -1, 0 and $+1$.

6.3.2 2^n Factorial Experiments and Fractions of 2^n Factorial Experiments

Factorial experiments in which all combinations of two levels of each of n factors are used, possibly with replications, are referred to as 2^n factorial experiments.

In many books discussing them, the factors are denoted by A, B, C, etc. and the individual runs are denoted by (1), a, b, ab, etc., where factor letters included in the run label are those at the higher or second level. This notation is convenient for discussing the theory of fractional factorial designs, but experimenters generally find it easier to use a table.

Ways of blocking such factorial experiments such that the analysis remains simple and the effects are not confounded with differences between blocks are given in books on design.

6.3.3 Fractional Factorial Designs

This section discusses fractional factorial designs in which trials are conducted at only a well-balanced subset of the possible combinations of levels of the factors.

A factorial design involving several factors would often require an extremely large number of trials and be uneconomic. It is often practical to perform what is called a fractional factorial experiment in which only a small fraction of the trials called for by the full factorial experiment are performed. This is particularly recommended when most of the quantities of interest can still be estimated using the smaller experiment.

Consider a 2^3 experiment involving three factors A, B and C. Geometrically, the $2^3 = 8$ trials in such an experiment are like the eight corners of a cube like that in Figure 5, in which the three directions of the edges correspond to the three factors.

If all eight trials in this experiment are performed then it will be possible to estimate:

- the overall mean;
- the main effect of factor A;
- the main effect of factor B;
- the main effect of factor C;
- the interaction of factors A and B;
- the interaction of factors A and C;
- the interaction of factors B and C, and;
- the three-factor interaction of factors A, B and C.

With eight data points we can estimate eight things.

If instead we performed only a carefully planned half of the factorial experiment we would still be able to estimate the overall mean and the main effects of the factors. This works best if we conduct runs corresponding either to the set of four corners of the cube which are shown as large blobs, or the complementary set of four runs. The four corners of the cube which are shown as large blobs form a balanced fractional factorial experiment because there are two of them on each face of the cube. Note that if one factor is irrelevant then the set of four trials gives a complete factorial experiment over the other two factors.

Table 2 gives this design as a matrix of values. If we regard the levels of the factors as being the numbers -1 and $+1$, then the runs in this design are the one for which the product of the levels of the three factors is -1.

DESIGN THE EXPERIMENT

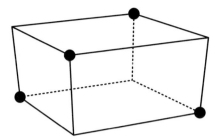

Figure 5. A cube as a representation of a 2^3 factorial experiment. The four corners plotted with large blobs constitute a balanced fractional factorial experiment

Table 2. Design for 3 factors at 2 levels in 4 runs

Standard Order	Factors A	B	C
1	−	−	−
2	−	+	+
3	+	−	+
4	+	+	−

There is a substantial body of theory about fractional factorial designs. For details, see any of a large number of books on experimental design such as Finney (1960), Davies (1967) or Box, Hunter and Hunter (1978). I will illustrate some of the capabilities of fractional factorial designs rather than discussing them thoroughly.

An example of a useful fractional factorial design with eight runs is given in Table 3. It enables the main effects of up to seven factors to be investigated, but there is a lot of confounding between main effects and two-factor interactions which could be of concern. For instance, factor D is confounded with the interaction between factors B and C.

6.3.4 Resolution

A concept which is important for fractions of 2^n factorial designs is the resolution of such designs. The formal definition of resolution is the minimum number of letters in equations defining the design. For the example in Table 3, the levels of factors A, B and C are in a standard order. The levels of the other factors could be calculated using the relationships $D = -BC$, $E = ABC$, $F = -AB$ and $G = -AC$, so the resolution is 3. Common practice is to use Roman numerals to describe resolution, so we would say that this design is one of resolution III.

Only resolutions III, IV and V are commonly important in practice.

Table 3. Design for 3–7 factors at 2 levels in 8 runs. For three factors, columns A, B and C give a factorial design. For four factors, columns A, B, C and E give a resolution IV design with interactions confounded in pairs: AB with CE, AC with BE and AE with BC. For five or six factors, designs which omit the last one or two columns are a reasonable choice. The designs with 5–7 factors are of resolution III, so a likely next experiment is the foldover design given in Table 4

Standard Order	Factors						
	A	B	C	D	E	F	G
1	−	−	−	−	−	−	−
2	−	−	+	+	+	−	+
3	−	+	−	+	+	+	−
4	−	+	+	−	−	+	+
5	+	−	−	−	+	+	+
6	+	−	+	+	−	+	−
7	+	+	−	+	−	−	+
8	+	+	+	−	+	−	−

Resolution III means that the main effects are confounded by two-factor interactions. This may be acceptable. If you are confident that there are no substantial interactions, then I recommend that you consider fractional factorial experiments of resolution III because fewer runs are often required than for one-factor-at-a-time experiments and the precision of estimating the effects is greater because all results are used for estimating all of the main effects.

Resolution IV means that main effects are not confounded by two-factor interactions. However two-factor interactions are confounded with one another.

If you would like to have some insurance against the possible presence of interactions, then I recommend fractional factorial experiments of resolution IV because they are much more reliable than one-factor-at-a-time experiments and fractional factorial experiments of resolution III, but they do not require as many runs as full factorial experiments.

Sometimes it is reasonable to not be concerned to estimate all interactions but to regard them as part of the noise – and it is essential to overcome noise so that a reproducible result can be obtained. For instance, it may be desirable that effects found experimentally in a laboratory can be reproduced during manufacturing or in a customer's environment. Interactions must be small relative to the main effects for this to be the case.

When used in such situations a resolution IV experiment should be followed up by a confirmatory experiment. If the responses from the confirmatory experiment accord with predictions, then this is generally adequate evidence that interactions are small relative to the main effects. If the confirmatory experiment does not agree with predictions, then there may be some substantial interactions and further information might need to be obtained before implementing the conclusions.

Resolution V or higher resolution means that two-factor interactions are not confounded with one another. This is the minimum useful resolution when it is

Table 4. Foldover of design in Table 3

Standard Order	Factors						
	A	B	C	D	E	F	G
1	+	+	+	+	+	+	+
2	+	+	−	−	−	+	−
3	+	−	+	−	−	−	+
4	+	−	−	+	+	−	−
5	−	+	+	+	−	−	−
6	−	+	−	−	+	−	+
7	−	−	+	−	+	+	−
8	−	−	−	+	−	+	+

desirable to estimate the magnitude of all two-factor interactions. Such a desire may be related to the results from a lower resolution experiment. However, lesser resolution may be acceptable if background information can be used to support the presumption that some interactions are negligible.

Sometimes later experiments must be run to resolve questions obscured by confounding. Sometimes a solution of appropriate reliability can be found without worrying about the uncertainty caused by confounding.

When experimenting with two-level factors, resolution III designs can always be made into resolution IV designs by doubling the number of runs. This was shown by Margolin (1969). The factor levels for the additional runs can often be derived by simply swapping all levels of some (commonly all) factors to be the opposite. This is referred to as a 'foldover' of the original design. For instance, Table 4 is a foldover of Table 3. Each of these experiments is an experiment of resolution III in seven factors. The combination of the experiments gives an experiment of resolution IV.

Two other useful small fractional factorial designs with factors at two levels are given in Tables 5 and 6. The design in Table 6 is referred to in Taguchi (1986) and Taguchi (1987) as L_{16}, but the columns are in a different order.

Example: Disassembly and reassembly

One way of searching for the principal causes of manufacturing problems such as noisiness in operation is to disassemble some products which have been observed to display those problems, and some products which were problem-free. A reassembly experiment could be conducted using a fractional factorial design. The factors are the components into which products have been disassembled. The possible levels of these factors are that the component came from a trouble-free product, or it came from a product which had displayed a problem.

Analysis of such an experiment will generally indicate which components are critical. Confounding of interactions is unlikely to be difficult to resolve, because it is expected that most of the factors will have no effect at all and few interactions are likely to have substantial influences.

Table 5. Plackett and Burman (1946) design for up to 11 factors at 2 levels in 12 runs. Two-factor interactions are partially confounded with all of the other main effects. The design which Taguchi refers to as L_{12} can be seen to be equivalent to this by reordering the columns into the order A, E, G, F, I, B, C, H, K, D, J and by reordering the rows into the order 1, 4, 8, 9, 10, 12, 11, 5, 6, 7, 2, 3

Standard Order	\multicolumn{11}{c}{Factors}
	A	B	C	D	E	F	G	H	I	J	K
1	+	+	−	+	+	+	−	−	−	+	−
2	+	−	+	+	+	−	−	−	+	−	+
3	−	+	+	+	−	−	−	+	−	+	+
4	+	+	+	−	−	−	+	−	+	+	−
5	+	+	−	−	−	+	−	+	+	−	+
6	+	−	−	−	+	−	+	+	−	+	+
7	−	−	−	+	−	+	+	−	+	+	+
8	−	−	+	−	+	+	−	+	+	+	−
9	−	+	−	+	+	−	+	+	+	−	−
10	+	−	+	+	−	+	+	+	−	−	−
11	−	+	+	−	+	+	+	−	−	−	+
12	−	−	−	−	−	−	−	−	−	−	−

Table 6. Design for 4–15 factors at 2 levels in 16 runs. For four factors, columns A, B, C and D give a factorial design. For five factors, columns A, B, C, D and I give a design of resolution V. For 6–8 factors, columns A, B, C, D, H=BCD, L=ACD, M=ABC and N=ABD give a design of resolution IV. For 9–15 factors, omit the last few columns for a useful design of resolution III and consider using a foldover design as the next experiment

Standard Order	A	B	C	D	E	F	G	H	I	J	K	L	M	N	O
1	−	−	−	−	−	−	−	−	−	−	−	−	−	−	−
2	−	−	−	+	−	−	+	+	+	−	+	+	−	+	+
3	−	−	+	−	−	+	−	+	+	+	−	+	+	−	+
4	−	−	+	+	−	+	+	−	−	+	+	−	+	+	−
5	−	+	−	−	+	−	−	+	+	+	+	−	+	+	−
6	−	+	−	+	+	−	+	−	−	+	−	+	+	−	+
7	−	+	+	−	+	+	−	−	−	−	+	+	−	+	+
8	−	+	+	+	+	+	+	+	−	−	−	−	−	−	−
9	+	−	−	−	+	+	−	+	−	−	+	+	+	−	−
10	+	−	−	+	+	+	−	+	−	−	+	−	+	−	+
11	+	−	+	−	+	−	+	+	−	+	−	−	−	+	+
12	+	−	+	+	+	−	−	−	+	+	+	+	−	−	−
13	+	+	−	−	−	+	+	+	−	+	+	+	−	−	−
14	+	+	−	+	−	+	−	−	+	+	−	−	−	+	+
15	+	+	+	−	−	−	+	−	+	−	−	+	+	−	+
16	+	+	+	+	−	−	−	+	−	−	−	+	+	+	−

DESIGN THE EXPERIMENT

Table 7. Design for 3–4 factors at 3 levels in 9 runs

Standard Order	A	B	C	D
1	−	−	−	−
2	−	0	0	0
3	−	+	+	+
4	0	−	0	+
5	0	0	+	−
6	0	+	−	0
7	+	−	+	0
8	+	0	−	+
9	+	+	0	−

6.3.5 3^n Factorial Experiments and Fractions

A factorial experiment in which each factor has three levels is called a 3^n factorial experiment. The theory of fractions of 3^n factorial experiments is more difficult than that of fractions of 2^n experiments. Details can be found in books. For instance, Cochran and Cox (Plan 6A.19 page 291) show how a one-third replicate of a 3^5 factorial can be run in blocks of nine units in such a way that all main effects and two factor interactions (except one) can be estimated. A useful small fractional factorial design with up to four factors at three levels is given in Table 7. This design is referred to in Taguchi (1986) and Taguchi (1987) as L_9.

Another example of such a design is given in Table 8. This design is referred to in Taguchi (1986) and Taguchi (1987) as L_{27}. It allows a large number of factors to be investigated at three levels.

6.3.6 Taguchi's Orthogonal Arrays

Genichi Taguchi was so successful at selling some ideas about experimenting that 'The Taguchi Approach' might be regarded as a well-known brand name. I agree with the opinion of Box (1985) about Taguchi's contribution:

> I believe that Taguchi's engineering ideas are important and novel and that we must understand and use them. But I also believe that it would be foolish to learn the details of many of the statistical methods that he has proposed.

Part of Taguchi's success at getting people to experiment was that Taguchi encouraged the use of a small number of experimental designs. These could then be readily used without experimenters needing to understand much theory in order to lay out the runs of an experiment. However, I believe that there is no justification for associating the term 'orthogonal arrays' with Taguchi's name. The term is best regarded as a synonym for 'fractional factorials'.

Table 8. Design for 5–13 factors at 3 levels in 27 runs

Standard Order	Factors												
	A	B	C	D	E	F	G	H	I	J	K	L	M
1	−	−	−	−	−	−	−	−	−	−	−	−	−
2	−	−	−	−	0	0	0	0	0	0	0	0	0
3	−	−	−	−	+	+	+	+	+	+	+	+	+
4	−	0	0	0	−	−	−	0	0	0	+	+	+
5	−	0	0	0	0	0	0	+	+	+	−	−	−
6	−	0	0	0	+	+	+	−	−	−	0	0	0
7	−	+	+	+	−	−	−	+	+	+	0	0	0
8	−	+	+	+	0	0	0	−	−	−	+	+	+
9	−	+	+	+	+	+	+	0	0	0	−	−	−
10	0	−	0	+	−	0	+	−	0	+	−	0	+
11	0	−	0	+	0	+	−	0	+	−	0	+	−
12	0	−	0	+	+	−	0	+	−	0	+	−	0
13	0	0	+	−	−	0	+	0	+	−	+	−	0
14	0	0	+	−	0	+	−	+	−	0	−	0	+
15	0	0	+	−	+	−	0	−	0	+	0	+	−
16	0	+	−	0	−	0	+	+	−	0	0	+	−
17	0	+	−	0	0	+	−	−	0	+	+	−	0
18	0	+	−	0	+	−	0	0	+	−	−	0	+
19	+	−	+	0	−	+	0	−	+	0	−	+	0
20	+	−	+	0	0	−	+	0	−	+	0	−	+
21	+	−	+	0	+	0	−	+	0	−	+	0	−
22	+	0	−	+	−	+	0	0	−	+	+	0	−
23	+	0	−	+	0	−	+	+	0	−	−	+	0
24	+	0	−	+	+	0	−	−	+	0	0	−	+
25	+	+	0	−	−	+	0	+	0	−	0	−	+
26	+	+	0	−	0	−	+	−	+	0	+	0	−
27	+	+	0	−	+	0	−	0	−	+	−	+	0

DESIGN THE EXPERIMENT

Table 9. Design for 0–1 factors at 2 levels and 2–7 factors at 3 levels in 18 runs. Note that factor A and B can be combined into a 6-level factor

Standard Order	A	B	C	D	E	F	G	H
1	−	−	−	−	−	−	−	−
2	−	−	0	0	0	0	0	0
3	−	−	+	+	+	+	+	+
4	−	0	−	−	0	0	+	+
5	−	0	0	0	+	+	−	−
6	−	0	+	+	−	−	0	0
7	−	+	−	0	−	+	0	+
8	−	+	0	+	0	−	+	−
9	−	+	+	−	+	0	−	0
10	+	−	−	+	+	0	0	−
11	+	−	0	−	−	+	+	0
12	+	−	+	0	0	−	−	+
13	+	0	−	0	+	−	+	0
14	+	0	0	+	−	0	−	+
15	+	0	+	−	0	+	0	−
16	+	+	−	+	0	+	−	0
17	+	+	0	−	+	−	0	+
18	+	+	+	0	−	0	+	−

6.3.7 Designs and Problems that Don't Seem to Fit

Fractional factorial designs are generally only available when either all factors have two levels or all factors have three levels. The designs which are best-known by Taguchi's names of L_{18} and L_{36} allow one two-level factor and up to 11 two-level factors to be used in combination with three-level factors which allow mixed numbers of levels.

Real problems are often stated in the first instance as requiring various numbers of levels for various factors. There are three useful tactics for using fractional factorial designs when the problem doesn't seem to fit the design.

- Sometimes decisions can be made to restrict attention to a number of levels that fits the available designs. This is commonly acceptable in a sequence of experiments because the effects of other levels can be estimated in subsequent experiments.
- If there is just one factor that has many levels then it is frequently possible to include such a factor in a fractional factorial design by combining two or more factors and their interactions into a single factor. For two-level factors, two columns can be combined to make a column for one four-level factor. This can be done by taking twice the first column plus the second column and relabelling the four distinct answers obtained as '1', '2', '3' and '4'. The column containing the interaction between the original columns must not be used for another factor because it would be confounded with the 4-level factor.

Table 10. Design for up to 11 factors with 2 levels (A–K) and up to 12 factors with 3 levels in 36 runs (L–W). Note that columns A–K can be combined into a 12-level factor

	A	B	C	D	E	F	G	H	I	J	K	L	M	N	O	P	Q	R	S	T	U	V	W
1	−	−	−	−	−	−	−	−	−	−	−	−	−	−	−	−	−	−	−	−	−	−	−
2	−	−	−	−	−	−	−	−	−	−	−	0	0	0	0	0	0	0	0	0	0	0	0
3	−	−	−	−	−	−	−	−	−	−	−	+	+	+	+	+	+	+	+	+	+	+	+
4	−	−	−	−	−	+	+	+	+	+	+	−	−	−	−	0	0	0	0	+	+	+	+
5	−	−	−	−	−	+	+	+	+	+	+	0	0	0	0	+	+	+	+	−	−	−	−
6	−	−	−	−	−	+	+	+	+	+	+	+	+	+	+	−	−	−	−	0	0	0	0
7	−	−	+	+	+	−	−	−	+	+	+	−	−	0	+	−	0	+	+	−	0	0	+
8	−	−	+	+	+	−	−	−	+	+	+	0	0	+	−	0	+	−	−	0	+	+	−
9	−	−	+	+	+	−	−	−	+	+	+	+	+	−	0	+	−	0	0	+	−	−	0
10	−	+	−	+	+	−	+	+	−	−	+	−	−	+	0	−	+	0	+	0	−	+	0
11	−	+	−	+	+	−	+	+	−	−	+	0	0	−	+	0	−	+	−	+	0	−	+
12	−	+	−	+	+	−	+	+	−	−	+	+	+	0	−	+	0	−	0	−	+	0	−
13	−	+	+	−	+	+	−	+	+	+	−	−	0	+	−	+	0	−	+	+	0	−	0
14	−	+	+	−	+	+	−	+	+	+	−	0	+	−	0	−	+	0	−	−	+	0	+
15	−	+	+	−	+	+	−	+	+	+	−	+	−	0	+	0	−	+	0	0	−	+	−
16	−	+	+	+	−	+	+	−	+	−	−	−	0	+	0	−	−	+	0	+	+	0	−
17	−	+	+	+	−	+	+	−	+	−	−	0	+	−	+	0	0	−	+	−	−	+	0
18	−	+	+	+	−	+	+	−	+	−	−	+	−	0	−	+	+	0	−	0	0	−	+
19	+	−	+	+	−	−	+	+	−	+	−	−	0	−	+	+	+	−	0	0	−	0	+
20	+	−	+	+	−	−	+	+	−	+	−	0	+	0	−	−	−	0	+	+	0	+	−
21	+	−	+	+	−	−	+	+	−	+	−	+	−	+	0	0	0	+	−	−	+	−	0
22	+	−	+	−	+	+	+	−	−	−	+	−	0	0	+	+	−	0	−	−	+	+	0
23	+	−	+	−	+	+	+	−	−	−	+	0	+	+	−	−	0	+	0	0	−	−	+
24	+	−	+	−	+	+	+	−	−	−	+	+	−	−	0	0	+	−	+	+	0	0	−
25	+	−	−	+	+	−	+	+	−	−	−	−	+	0	−	0	+	+	−	+	−	0	0
26	+	−	−	+	+	−	+	+	−	−	−	0	−	+	0	+	−	−	0	−	0	+	+
27	+	−	−	+	+	−	+	+	−	−	−	+	0	−	+	−	0	0	+	0	+	−	−
28	+	+	+	−	−	−	−	+	+	−	+	−	+	0	0	0	−	−	+	0	+	−	+
29	+	+	+	−	−	−	−	+	+	−	+	0	−	+	+	+	0	0	−	+	−	0	−
30	+	+	+	−	−	−	−	+	+	−	+	+	0	−	−	−	+	+	0	−	0	+	0
31	+	+	−	+	−	+	−	−	−	+	+	−	+	+	0	+	0	0	−	0	−	−	−
32	+	+	−	+	−	+	−	−	−	+	+	0	−	−	−	+	−	+	+	0	+	0	0
33	+	+	−	+	−	+	−	−	−	+	+	+	0	0	0	−	0	−	−	+	−	+	+
34	+	+	−	−	+	−	+	−	+	+	−	−	+	−	0	+	0	+	−	0	0	+	−
35	+	+	−	−	+	−	+	−	+	+	−	0	−	0	+	−	+	−	0	+	+	−	0
36	+	+	−	−	+	−	+	−	+	+	−	+	0	+	−	0	−	0	+	−	−	0	+

DESIGN THE EXPERIMENT

Similarly, two 3-level factors and the two columns which contain their interaction may be combined into a 9-level factor, and three 2-level factors and the four columns which contain all the interactions between them may be combined into one 8-level factor. These options are particularly likely to be useful if an experiment is to be conducted in blocks.

- If the number of formal levels of a factor in a design is too large, then some of the actual levels can be used more than once. For instance, if there are two machines to be tried in a three-level fractional factorial experiment then the formal levels of one factor can be taken to be 'Machine A', 'Machine A' and 'Machine B'.
The analysis of the experiment becomes slightly more complicated because such designs are unbalanced.

6.3.8 Designs Used for Response Surface Methodology

In this section we discuss types of experimental designs which are useful when the models to be fitted include interaction terms, quadratic terms, and possibly higher order terms also.

My starting point for choosing a design in such situations is to consider what is called a 'central composite' design. Such designs are constructed by adding centre points and star points to 2^n factorial designs or fractions.

- Centre points are points such that the level of all factors is at the mid-point of levels used in the factorial design. If the design is blocked then it is common practice to include a centre point in each block. Replication of the centre point gives information about repeatability where it is likely to be most relevant.
- Star points are points such that all factors except one are at the mid-point of levels used in the factorial design. That one factor generally has a level, often denoted by $\pm a$, outside the range used in the factorial part of the experiment.
- The position of the star points and the number of centre points are sometimes chosen so that the design is rotatable in the sense that it is equally informative about the response surface in all directions from the centre of the design. However, this is not particularly important.

One example of such a design in two factors is given in Table 11.

Another example of such a design is given in Table 12. Runs 1–4 and 6–9 constitute a factorial design; runs 11–16 are star points and runs 5, 10 and 17 are centre points. Figure 6 shows the geometry of this design.

Box–Behnken designs are a another useful class of designs for investigating response surfaces. See Box and Behnken (1960). Box and Draper (1987, section 7.5) give some worked examples and further references to this type of design. They recommend the use of repeated centre points (equally divided between blocks, where appropriate). Figure 7 shows the geometry of a Box–Behnken design with three factors.

Uniform shell designs are yet another way of specifying a set of points on the surface of a sphere in several dimensions. See Doehlert (1970) and Doehlert and Klee (1972) for details. A uniform shell design in two factors is given in Table 14. In three dimensions, the uniform shell design is equivalent to the Box–Behnken design.

When using a central composite, Box–Behnken and uniform shell designs, it is useful to regard them as consisting of points on the surface of a hypersphere (in some number

Table 11. Central composite design in two factors

Standard Order	Factors A	B
1	−1	−1
2	−1	+1
3	+1	−1
4	+1	+1
5	0	0
6	−α	0
7	+α	0
8	0	−α
9	0	+α
10	0	0

Table 12. Central composite design for 3 factors using a centre point in each of three blocks

Standard Order	Factors A	B	C
1	−1	−1	−1
2	−1	+1	+1
3	+1	−1	+1
4	+1	+1	−1
5	0	0	0
6	−1	−1	+1
7	−1	+1	−1
8	+1	−1	−1
9	+1	+1	+1
10	0	0	0
11	−α	0	0
12	+α	0	0
13	0	−α	0
14	0	+α	0
15	0	0	−α
16	0	0	+α
17	0	0	0

DESIGN THE EXPERIMENT

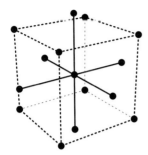 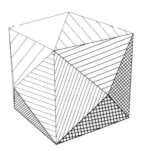

Figure 6. Two views of a central composite design in three dimensions. On the left, we see that it consists of a centre point (generally replicated), the eight points at the corners of a cube, and six points each displaced along one of the major axes from the centre. On the right, we see a polyhedron which looks like square prisms on the faces of a cube

Table 13. A Box–Behnken design with 3 factors, using a centre point in each of three blocks of five runs

Standard Order	Factors		
	A	B	C
1	−1	−1	0
2	−1	+1	0
3	+1	−1	0
4	+1	+1	0
5	0	0	0
6	−1	0	−1
7	−1	0	+1
8	+1	0	−1
9	+1	0	+1
10	0	0	0
11	0	−1	−1
12	0	−1	+1
13	0	+1	−1
14	0	+1	+1
15	0	0	0

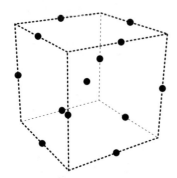 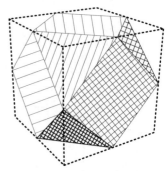

Figure 7. Two views of a Box–Behnken design. On the left, we see points at the middles of the sides of a cube, plus a centre point. On the right, we see a polyhedron which can be regarded as a cube with its corners cut off or as a octahedron with its corners cut off

Table 14. Uniform shell design in 2 factors

Standard Order	Factors	
	A	B
1	+0.866	−0.5
2	0	+1.
3	−0.866	−0.5
4	0	0
5	+0.866	+0.5
6	0	−1.
7	−0.866	+0.5
8	0	0

of dimensions) plus centre points. The first step in using a design is deciding how large a sphere you would like to investigate with your experiment. Usually, the scales and units on the various factors will be different. The spherical region will be transformed into an ellipsoid.

Why pay little attention to 'optimal' designs?

A substantial amount of research has been done on experimental designs which are D-optimal, A-optimal, E-optimal, etc. I suggest that most experimenters pay little attention to this body of research. I offer three reasons for this opinion.

- Experimental designs need to satisfy many criteria such as those listed below. I tend to distrust procedures which try too hard to optimize any single criterion.
  — It is desirable to minimize the number of experimental runs.
  — It is desirable that an estimate of error variance can be obtained.

- It is desirable to differentiate the replication error variance from the lack of fit of a quadratic surface. This requires either that the experiment includes replicate observations or that an estimate of replication error be obtainable from another source of information. It allows the goodness of fit of the quadratic surface to be checked.
- It is desirable to be able to check the assumption that the error variance is approximately constant over the region of interest.
- It is desirable to be able to check for outliers.
- It is desirable that a quadratic surface can still be fitted even if a small number of observations are missing for any reason.
- It is desirable that experiments can be performed in blocks.
- It is desirable that a first experiment which allows a linear model to be fitted can be used as the first block in a larger experiment which allows a quadratic surface to be fitted.

- The various optimality criteria which have been studied tend to suggest that widely spaced designs are always better than narrowly spaced designs. Yet the quadratic surfaces which are commonly fitted are likely to be a better fit over a narrow range than over a wide range. Since the criteria being optimized perform poorly on the decision of how widely spaced to take the points of a design, I tend to distrust other uses of the criteria also.
- I prefer to use designs on hyperspheres rather than designs on hypercubes, because I believe that the region of interest in practical situations is more likely to be better approximated by a hypersphere (after a scaling of the axes). See Lucas (1976) for a discussion of designs on hyperspheres and hypercubes.

As an example in two dimensions, suppose your best current estimate of the optimal operating conditions for some process is

Temperature: 200°C, Amount of X: 100 g

and suppose that you regard the options of changing the temperature by 5°C and leaving the amount of X unchanged, or changing the amount of X by 3 g and leaving the temperature unchanged as equally likely to be an improvement.

The question that you need to ask yourself is whether you think that the rectangle or the ellipse in Figure 8 better corresponds to your region of interest. These regions have the same area.

I claim that the ellipse is a more sensible region of interest than the rectangle. For instance, the conditions shown as point A, namely

Temperature: 205.3°C, Amount of X: 100 g

are nearer to the preliminary estimate of optimal operating conditions than are those at B, which are

Temperature: 204.5°C, Amount of X: 102.5 g.

(The definition of closeness is one which says that a difference of 5°C is just as large as a difference of 3 grams of X.)

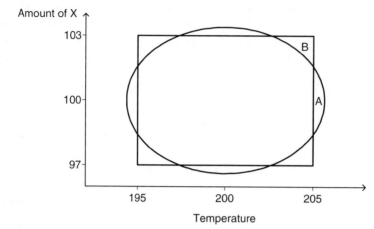

Figure 8. Two possible regions of interest for an experiment, a rectangle and an ellipse. They are the same size. In preferring the ellipse, I am judging that points like A are of greater interest than points like B because they are 'closer' to the middle of the region

6.3.9 Some Experimental Designs with other Possibly-useful Features

There are a very large number of experimental designs available. This section presents a few types of design in order to give an indication of the features which are available, so that you are more likely to consider looking for a fancy design for a particular situation.

6.3.10 A Staggered Nested Design

Batches of coarse (22 mm nominal topsize) iron ore regularly arrive at a laboratory for routine sample preparation and testing. Sample preparation involves several stages of crushing and sample reduction, and finishes with a container of powder which has a nominal topsize of 150 μm. Testing involves taking a sample of the 150 μm powder, preparing a bead and testing it in an X-ray fluorescence machine.

One way to estimate the bias and precision of the routine procedures would be for both the usual personnel and experts to prepare samples in duplicate and for all prepared samples to be tested in duplicate by both the usual personnel and experts. This might be done for, say, 20 batches. This is illustrated in Figure 9.

One problem with this design is that the operators know that the bias and precision of their procedures are being checked, so the results may not be typical of normal bias and precision. Another problem is that there is much more information about testing bias and precision than there is about sampling bias and precision.

An alternative design is for the usual personnel to prepare samples and test them in the usual way. (The only modification to normal procedures was that material which did not become part of the prepared samples was systematically collected.) Let us refer to the routine result on the ith sample as $X_{RR}(i)$. Experts then prepare samples in duplicate and to do all testing in duplicate. The experts also test the 150 μm powder resulting from the routine sample preparation, and do this testing in duplicate. Let

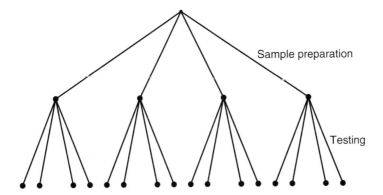

Figure 9. Diagram illustrating the structure of a nested design. The top point corresponds to an incoming sample. Sample preparation by experts is shown as a line downwards and to the right, while routine sample preparation is shown as a line to the left. Similarly, testing by experts is shown as a line to the right and routine testing is shown as a line to the right

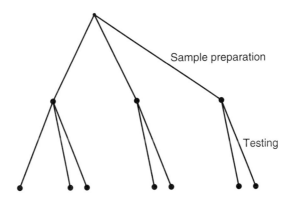

Figure 10. Diagram illustrating the structure of a staggered nested design. The design is the same as that shown in Figure 9 except that parts of it have been omitted

us refer to the experts' results on the ith sample as $X_{R1}(i)$, $X_{R2}(i)$, $X_{11}(i)$, $X_{12}(i)$, $X_{21}(i)$ and $X_{22}(i)$, where the first subscript is R, 1 or 2 according to whether the sample of 150 μm powder being testing was from routine preparation, the first expert preparation or the second expert preparation, and the second subscript indicates which of the duplicate tests is being recorded. This is illustrated in Figure 10.

Comparing the average $X_{RR}(i)$ with the average of the $X_{R1}(i)$ and $X_{R2}(i)$ indicates the bias of testing. The variability of the differences between duplicate tests by experts allows the variation due to expert testing to be estimated. The variability of the differences between duplicate sample preparations allows the variation due to expert sample preparation to be estimated. The variability in the values of $X_{RR}(i) - (X_{R1}(i) + X_{R2}(i))/2$ allows the precision of routine testing to be estimated,

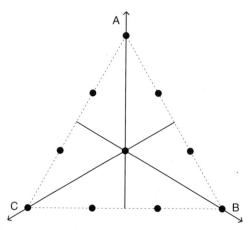

Figure 11. A simplex-lattice design for a three-component mixture experiment, shown as a set of ten points. The three axes labelled A, B and C show the directions of increasing amounts of three components in a mixture. The dotted lines delimit the region of possible mixtures, since the proportions must all be the range from zero to one

after subtracting the contribution due to variation in expert testing. Similarly, the bias and precision of routine sample preparation can be estimated.

This design is described as 'nested' because multiple tests can be done on a single prepared sample of 150 μm powder. It is described as 'staggered' because the amount of replication is not uniform over the design.

6.3.11 Designs for Experimenting with Mixtures

'Mixture experiments' is a term used to describe experiments where the controllable factors are the proportions of the components of a mixture. One of the unusual features of these designs is that the region of interest is a simplex, because of the constraint that the proportions must add to 100%. An example of a simplex-lattice design is shown in Figure 11.

There is a substantial literature on the analysis and design of mixture experiments. For instance, see Cornell (1981).

Many experiments involving mixtures are such that not all of the single-component mixtures are of interest, so the region of interest is not the entire simplex. In such circumstances, the most satisfactory way of incorporating background knowledge into the design of an experiment is often to vary the proportions of the minor components of the mixture so that the region of interest is covered, and to define the proportion of the major component as the discrepancy from 100%.

6.3.12 Latin Squares

Consider the following table of numbers, a Latin Square.

DESIGN THE EXPERIMENT

```
1 2 3 4
2 1 4 3
3 4 1 2
4 3 2 1
```

Each of the numbers 1, 2, 3 and 4 occurs once in each row and once in each column. Three factors each at four levels could be investigated in 16 trials by making one factor correspond to rows, one to columns and one to the number given in the table.

Except for 6 × 6 tables, a second Latin square orthogonal to the first can be found and used to investigate a fourth factor in the same experiment. Often, the numbers in the first table are coded as Latin letters and the numbers in the second table are coded as Greek letters – hence such designs are often referred to as Graeco-Latin Squares. They can be useful when there are two blocking factors in a single experiment.

Two further Latin Squares which are orthogonal to the Latin Square above and to each other are as follows. A useful reference is Fisher and Yates (1938). Larger sets of orthogonal Latin squares exist, but only when the numbers of rows and columns are larger and not for all sizes of square.

```
1 2 3 4        1 2 3 4
3 4 1 2        4 3 2 1
4 3 2 1        2 1 4 3
2 1 4 3        3 4 1 2
```

6.3.13 Balanced Incomplete Block Designs

When the number of trials which can be accommodated in a single block is small, a suitable design is often a balanced incomplete block design in which every pair of treatments occurs together in a block the same number of times as every other pair of treatments.

An example of such a design with four treatments in blocks of three is as follows. Each pair of treatments occurs together within a block two times.

Block 1: | A B C | Block 2: | A B D |
Block 3: | A C D | Block 4: | B C D |

6.3.14 Carryover Designs

These are designs in which every treatment is preceded by each other treatment the same number of times. Generally, each treatment is also preceded by no other treatment the same number of times. Each treatment is preceded by itself in some designs but not others, like the following.

124	235	346	457	561	672	713
142	253	364	475	516	627	731
154	265	376	417	521	632	743

A colleague, Richard Jarrett, suggested this design to me when I discussed a situation to him requiring the following features.

Table 15. Two possible orders for running a 2^3 factorial experiment

Standard order

Run	A	B	C
1	1	1	1
2	−1	1	1
3	1	−1	1
4	−1	−1	1
5	1	1	−1
6	−1	1	−1
7	1	−1	−1
8	−1	−1	−1

Suggested order

Run	A	B	C
1	1	1	1
2	−1	−1	−1
3	1	1	−1
4	−1	−1	1
5	1	−1	−1
6	−1	1	1
7	1	−1	1
8	−1	1	−1

- There were 21 cows.
- There were 7 treatments.
- Three treatments were to be applied sequentially to each cow.
- Each treatment follows each other treatment once.

6.3.15 Neighbour Designs

Consider the following sequence of trials:

$$(4)\ 1\ 2\ 3\ 4\ 5\ 3\ 1\ 5\ 2\ 4\ (1)$$

The end trials are performed but are only used as neighbouring trials, and are not used for statistical analysis in the same way as the other trials. For instance, five different varieties of pear trees could be used in this sequence. Twelve trees would need to be planted but the yield would only need to be measured on ten trees.

Each treatment has each other treatment as neighbour the same number of times, here precisely once. This may be an appropriate form of balance for some types of experiments.

6.3.16 Designs Robust against Autocorrelated Noise

For industrial trials which are conducted sequentially and for agricultural trials which are conducted on a one-dimensional array of experimental plots, it is becoming common to fit a model which considers trends with time, or trends across the series of plots. See Verbyla et al. (1999) and Steinberg (1988).

There has been a small amount of theoretical research on the design of experiments for use when such trends are expected and the fitting of such models is anticipated. For instance, Saunders and Eccleston (1992) showed that a 2^3 factorial experiment conducted in the suggested sequence shown in Table 15 will generally give much more precise estimates of the effects of the factors than will the same experiment conducted in a standard order.

For a narrow use of the term 'experimental design', these possible orders might not be regarded as distinct. People might describe them as being simply a single design

DESIGN THE EXPERIMENT

being run in two different orders. However, the difference between orders does affect the conduct of the experiment and may affect the precision of the conclusions.

6.4 DECIDE WHICH FACTORS TO INCLUDE AND AT WHAT LEVELS

Consider the factors involved and the following points.

- Some factors need to be included to enable conclusions to be inferred for a sufficiently wide population.
- Some factors may need to be included to model variations associated with measurement.
- When factors are continuously variable, generally use levels equally spaced or in a constant ratio.
- Only use two levels when uncertain about the direction of effect or whether the effect is zero. Use three, or possibly more, if you want to fit a response curve.
- Choose levels far enough apart for differences to be detectable and relevant.
- Is there a naturally defined control treatment which should be included?
- Standardize factors not included in particular trials.

This is the hardest concern: choose levels to get a desired range in the response.

6.4.1 Choosing Indicative Factors and Levels for those Factors

Often it is desirable to include indicative factors in experiments in order to check whether the conclusions of the experiment are valid over the range of that factor. One example is operators.

6.4.2 Choosing Factors for Main Effects Experiments

In much industrial experimentation there are large numbers of factors which might influence an outcome. It is important to consider as many control factors as possible because the loss of not including a factor is likely to be more important than the loss of being uncertain about the error variance of the measurements.

Fractional factorial designs are very useful in such circumstances. Taguchi's presentation of fractional factorial designs as orthogonal arrays provides a useful way of thinking about experiments when there are large numbers of factors. Plackett–Burman designs and the idea of folding over designs are also useful.

A procedure for deciding what factors to include in an experiment is the following, in three stages.

1. Have a brainstorming session to identify as many variables as possible that might influence the outcome of interest. Involve all the people who you think might conceivably have a contribution to make to the task of identifying such variables. In particular, involve the practical people with hands-on experience with the processes and measurement procedures. The usual rules for conducting brainstorming sessions, below, apply.
   - No criticism!
   - Try to build on other people's ideas.

- No questioning of opinions! Questions should only be posed in order to clarify meaning.
- All suggestions to be recorded in a way so that everyone can see the list.

2. A small number of people should then classify these variables in order to think about them. Issues to think about include the following.
   - Some variables are easy to set directly.
   - Sometimes you must choose between variables which can only be set indirectly but have a straightforward effect on the outcome of interest and variables which are easy to set but are likely to interact with other variables.
   - Some variables should be regarded as noise variables – things which you cannot control for theoretical, practical or economic reasons but which you would like to make the process insensitive to.
   - Some variables must be considered in an experiment because they will affect the experimental results, but are of no interest in practice. For instance, boredom or improvement with practice must be considered in the design of an experiment.
   - Some variables can be altered quickly, easily and cheaply. Adjustment of other variables may be quite impractical or very expensive.

   Some new variables may come to light during this process. Some other variables may be seen to be redundant. Think also about combinations of variables which are likely to interact.

3. It is now necessary to decide which variables to include in a programme of experimentation. This is best done by consensus.

6.4.3 Choosing Levels for Factors in Main Effects Experiments

Having chosen factors to be varied in an experiment, the next step is to choose the levels to be used for those factors. The first important question is whether to use two or three levels. Issues of relevance are as follows.

- If you are quite unsure about the direction of the effect of the factor then two levels are probably adequate.
- If the process has already been experimented on in the past then it is likely that some control factors have been set to local optima so three levels are called for.
- More than three levels are virtually never justified in factor screening experiments.

Where there is a currently accepted setting for a factor then it is often used as one of the levels in an experiment. When a factor has three levels then the currently accepted setting is generally used as the middle of these. When a factor is to only have two levels then the currently accepted setting may be used as one of these or the two levels may be displaced, one on each side of the currently accepted setting.

Selection of the range of the levels of a factor should be considered very carefully.

- Choose a range such that you expect to be able to detect an effect of the magnitude which you consider likely. In particular, if you suspect that a factor may be completely irrelevant then it is sensible to choose widely separated levels.
- If experimenting online then ranges for factors might be chosen conservatively.
- Note that if selecting levels seems very difficult then it is probably better to do a series of small experiments rather to start with a large experiment.

DESIGN THE EXPERIMENT

- Be very cautions if choosing levels so far apart that different mechanisms are likely to be operating. For instance, cloth could be dyed with a large amount of liquid compared to the amount of cloth or it could be dyed using a much smaller amount of liquid.

6.4.4 Choice of Factors and Levels for Factors in Response Surface Experiments

The following things must be considered.

- It is easier to detect the influence of a factor if the levels are chosen to be wide apart. This generally means that experimental levels are outside the usual working ranges for factors.
- Widely spaced factor levels may cause explosions or maintenance problems because a factor is set to levels outside the range originally envisaged. When experimenting on production equipment, unsaleable products may be manufactured.
- A more complicated response surface may be needed to fit what happens over a wider range. On the other hand, data far from the region of interest is of very little value.
- In the neighbourhood of an optimum, a response surface is not well approximated by using only linear terms. It is often better approximated by first- and second-order terms, so three levels is usually sufficient.
  Far from the optimum, first-order terms often suffice to show a direction in which improvement is likely to be found.
- If there is reason to suspect that a response surface has different curvature in some variable on the two sides of an optimum, then four levels for that factor may be justified.

6.4.5 Controls

For some types of experiments it is necessary or desirable to use a 'control'. There are two common reasons.

Sometimes an experiment has as its aim to test whether some treatment is better than no treatment at all, some basic treatment or some commonly accepted treatment. In such cases, the treatment (or lack of it) that the new treatment is to be compared with may be called a 'control'. The main point when analysing such experiments is to check whether the new treatment is better than the 'control'.

Sometimes a 'control' is used primarily to describe the conditions under which an experiment was conducted – rather like a matchbox is used in a photograph in order to indicate the conditions under which the photograph was taken. In such circumstances it may be appropriate to report the results for the 'control' but not use them in the formal statistical analysis. I believe that the following examples illustrate such sets of circumstances.

Example: Termites attacking plywood

The data for this example was provided by John Creffield and Nam-Ky Nguyen. A complete description of the experiment is given in Creffield and Nguyen (1988).

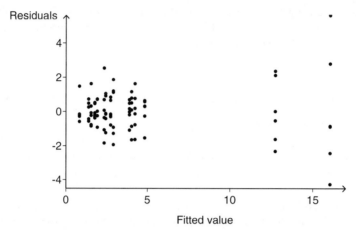

Figure 12. Plot of residuals against fitted values for an experiment in which pieces of wood were attacked by termites. The data for the two controls are on the right-hand side of the graph

The effectiveness of several preservation treatments for protecting plywood against attack by subterranean termite were compared. There were six pieces of plywood for each of 15 treatments, two of which were controls. The response variable was the mass loss of a piece of plywood after a standard amount of exposure to termites.

A model was fitted with each different treatment assumed to have a different mean response. Figure 12 is a plot of residuals against fitted values for that model. Note that on this figure there are sets of six points which have the same fitted value, being the average weight loss for the set of six pieces of plywood treated in a particular way.

The issue to be discussed is whether to include the results for the controls in the statistical analysis. I argue that those results should not be included.

- There is no doubt that the results for the controls are significantly different from the results for the other treatments. There is no need to include the controls so that a statistical test can be done.
- The variation amongst the results for the controls is not typical of the variability for the other treatments.
- The statistical analysis excluding the results for the controls has a smaller residual standard error. (It has a smaller R^2, but this is not important.)

Example: Testing insecticides

In an experiment testing the efficacy of several insecticides, one treatment was the spraying of water. The number of insects found after spraying the water gives an indication of the number of insects present. This is a matchbox-type control.

It would be a mistake to include the results for the control in the formal data analysis. Their variability is unrelated to the variability of the results for the insecticides.

6.5 CHOOSING A DESIGN

6.5.1 Catalogues of Designs

Catalogues of designs are available from a number of sources such as statistical computing packages. First, decide whether you want a main effects design or a design adequate for fitting a quadratic surface.

The most important steps in choosing a suitable design from a catalogue are to answer two questions. Namely, 'How many dimensions does the region of interest have?' and 'Is it important to restrict the number of different levels that some or all variables have?'.

- Do you need to estimate error variance from this experiment, or do you already have an estimate of adequate precision?
- Are systematic errors avoided? Would a very large experiment give the required information?
- Is the number of experimental units appropriate?
- Is the experiment simple enough? Do the people doing the experiment understand it? Are implementation and analysis feasible?
- Are all runs feasible? Are some potential results likely to be lost? What might go wrong?

The desirable type of experimental design is affected by purpose.

- Do you need information about the components of error variance such as the sampling and testing error? If this is the primary purpose of an experiment then you are likely to need a hierarchical design. If it is a secondary purpose then it will tend to increase the number of runs.
- Is this a factor screening experiment?
- Do you hope to fit a response surface model? If so, how complicated?

6.5.2 Think about What Might Go Wrong

What are the possibilities that something might go wrong?

- Are all runs feasible?
- Is the experiment simple enough?
- Can the design be analysed? Check this, using random numbers as data.
  There may be a mistake in a design, including those in this book. You may use a design incorrectly. (I have accidentally given a design to a client in which two factors were completely confounded, due to cross-referencing the same column of spreadsheet for both of the factors.)
- Might some runs be lost?

Simple experimental designs such as 2^3 and 3^2 factorials are suitable in a wide variety of situations. There is a lot of fancy theory available concerning experimental designs. However, a ten-year-old moderately priced car provides nearly as comfortable, safe and convenient transport as a new car.

6.5.3 Designs for Investigating Main Effects

In this section we discuss the types of designs, mostly fractional factorial designs, used when there are a large number of factors to be investigated at a relatively low precision.

Recommended useful designs are as follows.

- Simple comparative trials between two alternatives, usually a standard or normal way of running a process and a suggested alternative.
- One-factor experiments with a small number of levels being used for that factor.
- Simple factorial experiments, such as 2^2, 2^3, 2^4 and 3^2. Note that when running a 2^n factorial with continuously variable factors it is usually desirable to add an extra run at the centre of the factorial set of conditions.
- Plackett–Burman designs for screening large numbers of factors. Plan your resources to allow a 'foldover design' as outlined earlier to resolve ambiguities, if necessary.
- The half fractions of 2^3, 2^4 and 2^5 factorial experiments. These are useful exploratory designs. They can be followed by conducting runs constituting the remainder of the factorial design. So one way of thinking about using these designs is that you are planning to do a factorial experiment, but can analyse the experiment after half of the runs have been completed and possibly stop at that time.

6.5.4 Designs for Fitting Response Surfaces

Example with six factors

This example appears in Carter *et al.* (1983). Six factors were investigated using a laboratory simulator.

- Firing temperature was varied between 1080°C and 1380°C.
- Firing time was varied between 2 and 22 minutes.
- Alumina (Al_2O_3) content was varied between 2% and 4%.
- Silica (SiO_2) content was varied between 2% and 10%.
- Lime (CaO) content was varied between 0.1% and 1.3%.
- Ore fineness as measured by the Blaine index was varied between $200\,m^2kg^{-1}$ and $300\,m^2kg^{-1}$.

The design used was a central composite design of 60 experimental firings composed of half of a 2^6 (32 firings), four centre points and duplicates of star points with $\alpha = 0.8$ (24 firings). Four properties were measured on the pellets from each firing:

- abrasion index;
- cold compression strength;
- reducibility;
- porosity.

Response surfaces were fitted by multiple regression in the usual way.

The difficult part of handling a problem like this is to present the results of the analysis in a comprehensible way. Three techniques were used.

- Contours of a response, say cold compressive strength, can easily be plotted as a function of two factors, say temperature and Blaine index, for any particular level

DESIGN THE EXPERIMENT

of the other four factors. A 3 × 3 array of contour plots was draw for three levels of silica content and three levels of lime content.

Such an array of contour plots allows the effects of four variables to be displayed in a fairly clear fashion.

- Main effect plots normally show only the average effect of a single factor. Often it is sensible to plot two lines on a main effect plot, showing the effect of the factor for two particular levels of a second factor.
- Information can be displayed for a small number of likely chemical compositions of the ore (alumina, silica and lime percentages). Such graphs illustrate how the factors which are process control variables should be set in order to make acceptable pellets from particular ore types.

6.5.5 *Designs for Investigating Sampling and Testing Precision*

The designs most commonly used for checking sampling and testing procedures for bias and for estimating the various components of variability are hierarchical experiments which are like factorial experiments. Interlaboratory trials are sometimes conducted too early in the development of new test methods. The sensitivity of new test methods to parameters such as temperature and the purity of chemical reagents should be investigated first within a single laboratory.

6.6 CASE STUDY: INA TILE EXPERIMENTS

My information about this experiment has been taken from pages 79–83 of Taguchi (1986) and pages 399–423 of Taguchi (1987). This example illustrates several aspects of thinking about factors, levels and designs.

In the summer of 1953, Masao Ina of the Ina Seitō Company conducted a series of experiments. These were very influential in Japanese industry, because they illustrated the usefulness of experimentation.

Wall tiles are made by a process in which raw materials are pulverized and mixed in the desired proportions. This is actually done by pulverizing the stones, compounding with clays, wet-grinding in a ball mill, drying, pulverizing and then mixing with water – but the precise details are not relevant to the experimentation to be discussed. Then the tiles are moulded using a press, fired in a kiln, glazed and fired again in a kiln.

The tunnel kiln used for firing tiles has a substantial temperature gradient near the outside of the kiln because the burner is there. Tiles fired near the outside of the kiln tend to be defective in dimensions, shape and appearance, especially coloured tiles. In the past, it had been possible to avoid firing coloured tiles near the outside of the kiln by always placing white tiles there. Due to an increase in demand for coloured tiles it was decided to investigate the possibility of doing the first firing of coloured tiles near the outside of the kiln.

First experiment

The first experiment was a laboratory-scale experiment, using ball mills of 2 kg capacity rather than the 1 tonne capacity ball mill used in the actual production process. Seven tiles were made from each of 27 experimental mixtures of raw materials.

The background situation which I imagine influenced decision-making about factors and levels was that the processes of making tiles had been investigated previously. The experimenters were hoping to find a set of conditions which would allow satisfactory coloured tiles to be produced cheaply even if their first firing was near the outside of the kiln, but not expecting to find such a set of conditions. A highly probable outcome was that experimental results would be used to work out an economic trade-off between costs of production and quality of tiles.

In order to maximize the chance of finding a new set of conditions which would make the process more robust, it was desirable to include many factors in the experiment. Because a substantial amount of information was already available about the process and the process had probably been roughly optimized at some time in the past, it was sensible to use three levels rather than two for the factors. Perhaps some of the factors might have non linear effects which were not previously known to a great enough precision.

There might be interactions between factors, but these were not investigated in the first experiment. A rule for adjusting the amounts of agalmatolite and chamotte according to the amount of feldspar was available from past experience.

The levels chosen for factors were based on ranges which were known from past experience to be reasonable. The levels are not too variable, so that all levels of all factors were expected to allow tiles to be produced and measured. And the levels are variable enough that we can expect to be able to detect differences between the levels in an experiment of the size which was planned.

Running the experiment over three weeks provided a check that any conclusions drawn were likely to be generally applicable. Given the variation in the performance of the kiln from week to week, an experiment done over a short time period would have been considered to be unreliable.

Factors and levels used in the experiment were as follows.

- Clay type: either a more expensive, better clay (Gainome); a cheaper, inferior clay (Kibushi); or a 50–50 mixture.
- Feldspar type: either of two types of feldspar (Mikumo or Mihanayama) or a 50–50 mixture.
- Agalmatolite type: three brands of agalmatolite were considered (Mitsuishi, Shokozan and Tanigawa). Mixtures of these were also considered with agalmatolite type being a 9-level factor, the levels being the three simple brands, the three possible 50–50 mixtures, and the three possible 50–25–25 mixtures.

Taguchi's statistical analysis of the results did not take the structure of this 9-level factor into account but treated it as if the levels were nine unrelated varieties of wheat.

It is common that using too many levels is silly. Here, nine levels are used because it is easy to construct a design in which most factors have three levels and one factor has nine levels.

- Percentage of feldspar: levels used were 0, 2.5% and 5%.
- Percentage of agalmatolite: the usual amount of agalmatolite is 50% plus twice the amount of feldspar. The agalmatolite percentages were taken to be the usual amount and the usual amount plus or minus 2.5%. Hence the agalmatolite percentages were: 47.5%, 50% and 52.5% for 0% feldspar; 52.5%, 55% and 57.5% for 2.5% feldspar; and 57.5%, 60% and 62.5% for 5% feldspar.

DESIGN THE EXPERIMENT

- Percentage of chamotte (crushed defective product): the usual amount of chamotte is 7.5% minus the amount of feldspar. The chamotte percentages were taken to be the usual amount and the usual amount plus or minus 2.5%. Hence the chamotte percentages were: 5%, 7.5% and 10% for 0% feldspar; 2.5%, 5% and 7.5% for 2.5% feldspar; and 0%, 2.5% and 5% for 5% feldspar.
- Percentage of lime: levels used were 0, 2.5% and 5%.
- Time of firing: the first firings ('biscuit processing') were conducted at three different times, at weekly intervals, so that the results were more likely to be indicative of the range of results likely to be achieved. The kiln does not always behave in the same way.
- Whether first, second or third batch processed in the particular 2 kg capacity mill.
- Position within the kiln: inside top, inside 2nd level, inside 3rd level, inside bottom, outside top, outside middle, outside bottom. Note that only 'biscuit processing' was performed at the position indicated. The second firing (after glazing) was done in about the same position near the middle of the kiln for all tiles.

The quantity of clay was not used as a factor. It was fixed at 20% on the basis of previous experience. The quantity of pottery stone was also not used as a factor. It is determined by subtracting the percentages of clay, agalmatolite, feldspar and chamotte from 100%. Amounts of water and lime were not considered when calculating percentages. The amount of lime is measured as a percentage of the total quantity of clay, agalmatolite, feldspar, chamotte and pottery stone.

The design used was a split-plot design with the split-plots being the seven positions within the kiln and the main-plots being a 27-run fractional factorial design, like that shown in Table 8 on page 162.

Characteristics of the tiles which were analysed were as follows.

- Size calculated as the average length of the four sides of tiles.
- Squareness calculated as the range of the lengths of the four sides.
- Warpage measured as the displacement of the centre of the tile relative to the average of the two diagonals.
- Apparent flatness assessed subjectively on a scale from 0 to 5.
- Crazing assessed on a scale of 1, 2 or 3 by noting when fine cracks appear in the glaze when tiles are exposed to superheated steam in an autoclave.

The results of the analysis will not be presented here in detail. Major conclusions were as follows.

- Size: with the use of more lime, the size of the tiles increases and the difference in size between the inside and outside of the furnace decreases considerably.
- Squareness: tiles from the outside of the furnace were less square. The type of agalmatolite was also influential with Tanigawa being poor.
- Warpage: Figures 13, 14 and 15 show the estimated effects of the various factors. The upper graphs or numbers give the effects on the tiles at the outside of the kiln. The lower graphs or numbers give the effects on the tiles at the inside of the kiln.
  - Gainome is better clay than Kibushi.
  - Lime has little influence in the inside of the furnace but more lime reduces warpage on the outside.

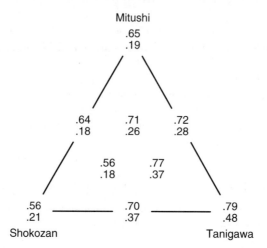

Figure 13. Diagram showing how the average warpage was influenced by the mixture of agalmatolite types. The upper of each pair of numbers gives the average warpage at the outside of the kiln, and the lower refers to the inside of the kiln. The positions of the pairs of numbers on the triangle indicate the mixture of agalmatolite types, with the proportion of, say, Mitushi, ranging from 100% at the vertex labelled Mitushi to 0% at the opposite side of the triangle

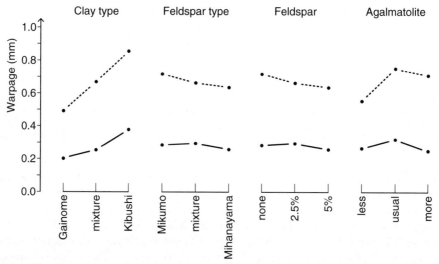

Figure 14. Main effects of factors clay type, feldspar type, feldspar amount and agalmatolite amount on the warpage. The upper, dashed lines give the average warpage at the outside of the kiln, and the lower, solid lines refer to the inside of the kiln

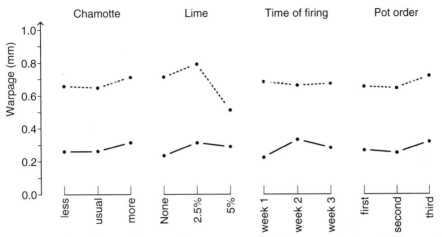

Figure 15. Main effects of factors chamotte amount, lime amount, time of firing and pot order on the warpage. The upper, dashed lines give the average warpage at the outside of the kiln, and the lower, solid lines refer to the inside of the kiln

— It is better to reduce the proportion of agalmatolite.
— Tanigawa is clearly the worst type of agalmatolite.

- Flatness:

  — Particularly at the outside of the furnace, Gainome is better clay than Kibushi.
  — High or low quantities of agalmatolite seem better than the intermediate quantity.
  — The less chamotte the better.
  — Tanigawa is the worst type of agalmatolite. Shokozan is somewhat better than Mitsuishi.
  — Using 2.5% lime is deleterious, but using 5% seems good, especially for tiles on the outside of the furnace.

- Crazing: only tiles from two positions on the outside of the kiln (2nd level and bottom) were tested because the testing was slow and destructive. The amount of lime had a large influence. It became apparent that this was because it increased the strength of the tiles. There was little difference between using 2.5% and 5% lime. Effects of clay type, agalmatolite type and agalmatolite percentage were observed but the preferred levels for these factors to reduce crazing conflicted with the levels which would be selected to optimize flatness and warpage.
- Uniformity of shape was also analysed, and the dimensions of the tiles were more uniform when more lime was used.

It was decided that it was now appropriate to run a second experiment at production scale.

Second experiment

The standard quantity of material mixed in a 'one-tonne' mill was 1200 kg. It was decided that about 1000 tiles would be fired in the kiln, 500 at inside positions and 500 at outside positions. Material in excess of that required to make 1000 tiles would be slowly blended with normal production material which was to be fired on the inner region of the kiln.

The full-scale experiment had an experimental design which was a fractional factorial design with eight runs, like that shown in Table 3 on page 158. A foldover of a further eight trials, like those shown in Table 4 on page 159, were to be run if the analysis of the first eight trials indicated that interactions between the quantity of lime and other factors needed to be estimated.

Factors and levels used were as follows.

- Clay type was fixed as Kibushi, price being taken to be of crucial importance for the time being. If some other way to improve quality can be found, then it is likely to be cheaper.
- Feldspar type was fixed as Mihanayama, since it seemed to have little influence.
- Agalmatolite type: either the current production mixture of mainly Mitsuishi and Shokozan or a test mixture with Tanigawa increased. As with clay type, the choice here was to reduce costs on the grounds that other methods of improving quality would be likely to be more cost-effective.
- Quantity of lime: either 1% (the current production level) or 5% as suggested by the laboratory-scale experimental results.
- Particle size of lime: either coarse (as currently) or fine.
  This factor was introduced because a new brand of lime had become available. Normally new factors would not be introduced in a full-scale experiment. However, trying a different brand seemed to be worth trying given that use of 5% lime is the most promising new option. Background knowledge and experience would have been adequate to suggest that the new factor is unlikely to have a catastrophic effect.
- Agalmatolite quantity: either 53% (current production level) or 43%.
- Mill charge quantity: either 1200 kg (current practice) or 1300 kg. To increase the charge quantity would reduce costs, so it would be desirable if the effect on product quality is not particularly adverse. Note that this factor could not have been investigated in the first experiment because it does not make sense at the laboratory scale.
- Chamotte quantity: either 0% or 4%.
- Feldspar quantity: either 0% or 5%.

The summary measurement made on each set of about 500 tiles was their monetary value assessed after grading according to normal inspection. This was expressed as an equivalent number of grade 1 tiles.

Analysis of the results indicated that to use 5% lime substantially improves the equivalent number of grade 1 tiles produced. This was the major conclusion in terms of its economic impact.

The better percentage of agalmatolite was 53% (the current level). The better mixture of types of agalmatolite was the present one used. The better mill charge quantity was 1200 kg (current practice). The better quantity of chamotte was 0%.

DESIGN THE EXPERIMENT

The better quantity of feldspar was 5%. The particle size of the lime did not have a statistically significant effect.

Next experiment

The next experiment conducted was to try the combination of factors suggested by the second experiment in full-scale production. The possible set of runs like the design in Table 4 on page 159 seem to never have been conducted.

Points illustrated by the Ina tile experiment

A major lesson which is supported by this sequence of experiments is that selections of factors and levels often have the most crucial influences on experimental outcomes.

With hindsight, the only issue that really mattered was to try using 5% lime. However, it is not fair to judge experimental designs in hindsight if you make use of information which you did not have before the experiment. Therefore, we should say that the only issue that really mattered was the exploratory attitude adopted in the selection of many factors at widely spaced levels in order to maximize the chance of getting a radical, new solution. Perhaps an experiment which included more factors at only two levels would have been better?

Usually in experiments with mixtures, the raw material which constitutes the largest proportion of the mixture is used to make the total to 100% and not used as a factor. Here, agalmatolite quantity was used as a factor because it was to be systematically varied with the amount of feldspar.

To make the levels of one factor depend on the levels of another factor may be very useful in removing interactions. It may also create interactions which would not otherwise have existed. It is not recommended unless there are reasons which can be advanced for it, either theoretical or based on experience.

Instead of a sequence within the pot mill, the identification of the pot mill could have been used as a factor. Neither would have been used except that there was a column of the orthogonal array available.

The only randomization done was to randomize the assignment of formal levels to factors. The experiment was conducted in standard order. The failure to randomize might not have been important, but for this experiment the expense of randomization would have been very small so I believe that it should have been done as a form of insurance.

Note that the advantages of sequential experimentation are exploited. Some criticisms of Taguchi suggest that he neglects this. The issues of 'siting of experiments' are well illustrated here. Note the emphasis on reducing costs, only the most cost-effective ways to improve quality are adopted.

Further details are available in Section 17.1 of Volume 1 of Taguchi's 'System of Experimental Design'. This experiment is of historical interest and it is one of the few for which most of the practical details are available.

It illustrates the point that many factors can be investigated in a quite small experiment, and that this is desirable if possibly useful factors are not to be overlooked. Here the major conclusion was that lime is useful for reducing sensitivity to firing temperature and the feature which distinguishes successful from unsuccessful possible experiments is whether or not lime percentage was considered as a factor. Taguchi's suggested strategy of using all available degrees of freedom to include additional factors reduces the possibility of missing useful factors.

Note that I have had to make some guesses about the state of knowledge of the people designing the experimental programme. I have not attempted to look for alternative descriptions of the experiment (which would probably be in Japanese) from the description given by Taguchi.

It is often the case that the full details of the state of knowledge of people designing an experimental programme are not available for reasons of commercial confidentiality or personal sensitivity.

7
Collect the Data

This chapter deals with the processes associated with getting experimental data after the experiment has been designed. These are:
- setting up a data-recording system;
- anticipating possible problems;
- performing runs, and;
- recording what happens.

7.1 SET UP A DATA-RECORDING SYSTEM

The design of a data-recording system could be considered to be part of the experimental protocol. Theoretically, it seems logical to me to classify data-recording as being part of experimental protocol. However, in practice, most experimenters make decisions about other aspects of experimental protocol much earlier than they make decisions about data-recording, so this topic is being dealt with separately.

Before designing your data-recording system, check that other aspects of your experimental protocol are explicitly defined. For substantial programmes of experimentation, the experimental protocol should be documented. Check that people know what is required for all steps in the conduct of the experiment. Will they record that these steps have been followed?

Try to minimize recording errors and to ensure that queries can be traced back to what actually happened.

- Record the intended levels of all factors, either explicitly or by using codenames for runs.
- Ensure that terminology and measuring equipment are standardized, in order to avoid problems.
- Order your recording system in such a way as to minimize likely errors. This generally means recording results of randomized trials in the randomized order, not in standard order. It also means minimizing manual transcription, possibly by using a photocopier.
- If samples or equipment must be labelled, regard this as a critical activity and organize a process which will not allow errors.
- Record any comments and the sequence number or other label of each run and the time when it was performed. These may be useful if something goes wrong and part of the experiment must be repeated.

7.1.1 General Remarks about Recording Data

Organize the recording of data in a way which makes accurate recording as likely as possible. Recording of data should be done as part of making the measurement. Where there is a choice between a manner of recording which makes recording easy and one which makes analysis easy always choose the manner of recording which makes recording easy. Use prepared data collection sheets or automatic equipment. Transcription of data for analysis can be checked at a later date, but errors in initial recording of results can never be rectified. Use unambiguous tags on samples.

As you perform an experiment you must record the data in a sensible way. This includes the following features.

- Record the intended levels of all factors.
- Ensure that terminology and measuring equipment are standardized, in order to avoid problems.
- Order your recording system in such a way as to minimize likely errors. This generally means recording results of randomized trials in the randomized order, not in standard order. It also means minimizing manual transcription, possibly by using a photocopier.
- If samples or equipment must be labelled, regard this as a critical activity and organize a process which will not allow errors.
- Try to avoid departures from intended levels of factors, but record any that occur.
- Record the sequence number or other label of each run and the time when it was performed. These may be useful if something goes wrong and part of the experiment must be repeated.
- Record all variables to be measured for each run.
- Record date, time of day, name of person conducting trial, identifier used for particular run, or any other information which you consider might be useful for minimizing the effects of a possible disaster.
- Record any modifications from the intended experimental protocol, including reasons why runs were not satisfactorily completed.
- Do not round off numerical results as part of the data-recording operation.
- Note any calibration of measuring equipment, measurements of quality of raw material or other information which might prove relevant when a later experiment suggests something as yet unforeseen.
- Check the reasonableness of the results as soon as possible after collecting them – suggestions that there have been minor departures from intended experimental procedures are easier to check the quicker you get onto them. Simple graphical methods are useful here.

Sometimes data management is a trivial exercise, but many projects would proceed more smoothly if a little time was spent thinking about how the data should be managed.

If you are using a computer system to manage your data, do not assume 'it's in the computer, it's okay!' Better and more realistic to assume that Murphy's Law holds.

Taking photographs is often a valuable precaution. I recommend taking colour slides because they have good resolution and are easy to file. Remember to have an identification system so that photographs are uniquely associated with experimental runs.

COLLECT THE DATA

Sometimes errors in data can be corrected unambiguously by referring to photographs. At other times, photographs allow you to look at features of what happened even though you did not know to look for those features at the time.

7.2 USING SPREADSHEETS FOR DATA RECORDING

This section discusses how spreadsheet computer programs can be used for data recording.

Many experimenters use computer spreadsheets for data recording. I believe that writing onto paper is a better option for the two reasons given below, but because spreadsheets are widespread I will give some opinions about how to use them most effectively.

The two reasons why I believe that writing experimental results onto paper is a better option than using a computer spreadsheet are as follows.

First when data is recorded onto paper and a correction is made for some reason then the old information can be crossed out without making it illegible and the new information can be recorded. Spreadsheets do not readily allow corrections to be made without destroying the old information.

In principle, spreadsheets can allow for corrections. Comments can be recorded. Old versions of spreadsheets can be kept and the computer operating system's information about the time when a file was last modified can be used to keep track of the versions. However, in practice, most people fail to keep a tidy record of anything other than the most recent version of the data.

I would like to see a computer software program designed for data recording. Such a program should allow some fields to be declared as always time-stamped, so that any entry was automatically associated with a record of when it was written and that even if such an entry were overwritten it could be viewed in some way. It would not be possible to delete the most recent entry and revert to an earlier one, but a copy of any early entry could be added to become the most recent.

The other basic problem with spreadsheets as a data-recording tool is that they are essentially rectangular. In the context of recording experimental data, this means that people tend to record the same number of data fields for each experimental run. This can be inappropriate if some parts of the experimental protocol are not done for each run, but are done either more or less often. The example of a spreadsheet for a bauxite digestion experiment given below provides an illustration of this problem. In order to get around the problem of needing to record the same number of data fields for each experimental run, people often make several copies of experimental data so that they are available wherever they are needed. This can create problems when data is corrected, updated or altered, because it is easy to fail to make a change in all of the places where the change should be made.

A computer-based data-recording tool could encourage users to recognise the hierarchical structure of data by, for instance, allowing for different types of unit records to be used in different windows. Such a data-recording tool might be used for a road accident study in the following way.

- One window would be used for recording information which was recorded only once for each accident – such as the location of the accident, the weather conditions, the date and the time of day.
- A second window would be used for recording information which will be recorded once for each vehicle involved in the accident – such as the type of vehicle, date of vehicle manufacture, contact details relevant to any brake inspection in previous 12 months and age of the driver of the vehicle.
- A third window would be used for information collected once for each vehicle occupant – such as age, whether sent to hospital, seating position in the vehicle and whether wearing a seatbelt.
- A fourth window would be used for information about particular injuries – such as the nature of the injury and the name of the person who assessed the severity of the injury.

If you do use a spreadsheet for recording data from an experiment then I have the following suggestions.

- Use different columns for the various measurements and characteristics, and use a separate row for each experimental run. This has the advantages that numbers which need to be compared are in the same column (people are much better at comparing numbers in columns than at comparing numbers in rows) and that data can be more conveniently transferred to a statistical analysis package or database program should this be required.
- If randomization of the order or records is done, then keep information such as the original record sequence number so that the randomization can be undone.
- Include calculations required for setting up the experimental runs as part of the same spreadsheet used for recording the results. It may be desirable to include copies of labels used for distinguishing samples or runs, with one copy near to the fields describing how to set up the runs and another copy near to where the results are to be recorded.
- Include calculations which check the reasonableness of results. This is the simplest way of ensuring that reasonableness checks will actually be done.
- Distinguish between fields used for different purposes. If using a colour monitor, I suggest that fields containing experimental data be blue, fields describing how the experimental runs should be set up be green, identifiers for runs or samples be black on on light yellow background, fields which monitor the reasonableness of the results should be red, and fields used for calculations be black. If using a black and white monitor, I suggest that fields containing experimental data be bold italic, fields describing how the experimental runs should be set up be bold, identifiers for runs or samples be bold white on on a black background, fields which should be monitored to check the reasonableness of the results should be bold italic white on on a black background, and fields used for calculations be in ordinary black lettering.
- Use generous amounts of space for headings. Include the measurement unit as part of each heading. Ensure that you give enough information so that each field is unlikely to be misinterpreted by anyone.
    — If a measurement can be made in more than one way then say which method was used.

COLLECT THE DATA

Table 1. Information recorded once for the entire experiment

Lime % (index)	Al2O3 %	SiO2 %	Fe2O3 %	TiO2 %	CaO %	LOI %	Qtz %
0	54.9	5.4	13.0	2.7	0	24.1	1.2
1	54.36	5.35	12.87	2.67	0.99	23.86	1.19
2	53.82	5.29	12.75	2.65	1.96	23.63	1.18

— If a measurement could be made at various times or on various samples then make clear exactly what time or sample has been measured.

If it is likely to be necessary to transfer data to a statistical analysis program then have an additional row for the abbreviated variable names which will be used in the statistical analysis program.
- Provide space in your spreadsheet for the date and time of performing operations or doing testing.
- Provide space for comments and encourage people to enter anything that might conceivably have any relevance whatsoever.
- Ensure that the computer files containing spreadsheets used for experimental data are backed up in an acceptable fashion.

7.2.1 Example: Spreadsheet for a Bauxite Digestion Experiment

For an experiment on the digestion of bauxite as part of the process of making alumina three worksheets were used.

1. The first sheet contained information measured or calculated only once for the entire experiment. A portion of this sheet is displayed in Table 1. The raw data described the chemical analysis of the bauxite, namely the percentages of Al_2O_3, SiO_2, Fe_2O_3, TiO_2, CaO_3, quartz and loss of moisture on ignition. The chemical analysis and measures of theoretically extractable alumina were also calculated for the addition of 1% or 2% of lime to the bauxite.
   The first column is the percentage of lime (as CaO) which is added to the bauxite in some cases. It is occasionally used an an index for looking up other information in this table. The sixth column gives the estimated percentage of CaO in the mixture. For instance, if 2% lime is added then the percentage of CaO is $2\%/102\% = 1.96\%$.
2. The second sheet contained information measured or calculated for the various batches of Bayer liquor which were used in the experiment. These batches of Bayer liquor were intended to have various levels of starting caustic concentration and concentrations of organic carbon. A portion of this sheet is displayed in Table 2. The actual concentrations of caustic and alumina were measured. The alumina-to-caustic ratio was calculated. It is a ratio that people who work in this area often use. Because of concern that the Bayer liquors may have changed over time, the concentrations of caustic and alumina for some of the liquors were retested and the changes in these concentrations and in the alumina-to-caustic ratio were monitored.

Table 2. Information recorded once for each batch of Bayer liquor

Block: Batch of Bayer liquor	Random numbers for blocks	Order of making up liquors	Design vars.		Starting conditions	
			S	O	caustic (g/L)	Organics (g/L)
1	0.4697	4	−1	−1	250	10
2	0.5993	8	−1	0	250	20
3	0.7429	10	−1	1	250	30
4	0.5324	5	0	−1	300	10
5	0.5583	6	0	0	300	20
6	0.4549	3	0	0	300	20
7	0.2511	2	0	1	300	30
8	0.5811	7	1	−1	350	10
9	0.1680	1	1	0	350	20
10	0.7335	9	1	1	350	30
11		11	2	0	400	20

The eleventh batch of Bayer liquor was not part of the original experimental design. No random number was generated for it.

The sixth and seventh columns which are used for setting up the experimental conditions were calculated from the design variables in the preceding two columns. For batches of Bayer liquor which were used for more experimental runs than could be conducted in a single day, the concentrations of caustic and alumina were measured at the time of the last usage of liquor from the batch. The changes in concentrations and in the alumina-to-caustic ratio were monitored.

3. The third sheet included calculation of the intended bauxite charge for each run according to a model which was updated after some of the runs had been done, and the actual bauxite charge used. After the dissolution of the bauxite in the Bayer liquor the concentrations of caustic, alumina and carbon in the resultant liquor were determined. Some of this sheet is shown as Tables 3 and 4. The undissolved and precipitated material, which is referred to as 'red mud' was chemically analysed for Al_2O_3, SiO_2, Fe_2O_3, TiO_2, Na_2O, CaO_3, quartz and loss of moisture on ignition. Quantities which were monitored to check the reasonableness of the data included the discrepancy between the total of the measured components of the red mud and 100%, the iron-to-titanium ratio in the red mud and ratio of sodium-to-(silica minus quartz) in the red mud. Various measures of the efficiency of the dissolution process were calculated, using numbers from all three of the worksheets.

7.3 TRY TO ANTICIPATE POSSIBLE PROBLEMS

Consider what might go wrong and whether anything be done to alleviate the subsequent difficulties? Ensure adequacy of raw materials, equipment and manpower before commencing.

COLLECT THE DATA

Table 3. First items of information recorded for each run

Std. order	Block number	Within block random numbers	Random numbers for blocks	Formal design					
				T	H	S	O	C	L
RunID	Block	RunRand	BlockRand	T	H	S	O	C	L
15	9	0.2243	0.1680	0	1	1	0	−1	0
47	9	0.4393	0.1680	1	0	1	0	0	−1
44	9	0.4705	0.1680	−1	0	1	0	0	1
11	9	0.5392	0.1680	0	−1	1	0	−1	0
12	9	0.7371	0.1680	0	−1	1	0	1	0
48	9	0.7663	0.1680	1	0	1	0	0	1
43	9	0.7824	0.1680	−1	0	1	0	0	−1
16	9	0.9212	0.1680	0	1	1	0	1	0
18	1	0.3152	0.4697	0	0	−1	−1	0	1
17	1	0.8383	0.4697	0	0	−1	−1	0	−1

Table 4. Further items of information recorded for each run

Set up of runs given liquor		Intended liquor			
Holding time (min) HoldTime	Lime (as % charge) LimeAdd	Starting caustic (g/L) IntCaustic	Organic carbon (g/L) IntOrgC	Actual starting caustic StCaustic	Actual starting A/C StA2C
7.5	1	350	20	337.4	0.3909
2.5	0	350	20	337.4	0.3909
7.5	2	350	20	337.4	0.3909
2.5	1	350	20	337.4	0.3909

Check the feasibility of a proposed experiment with the practical people who know about the likely problems. Think about the things that might go wrong – Murphy's Law is sure to apply if you don't!

- Check that raw materials will not run out or change in characteristics during the experiment.
- Is it necessary to run the process at new settings for a little while before taking samples or making measurements?
- Are the resources adequate?
- Is accurate measuring equipment available?
- Are skilled staff available?
- Is time available for experimental runs, or are the pressures of production likely to reduce this time?
- Has the likely real cost of the experiment been estimated and approved?

- Might the weather interfere with some experimental runs?
- What will happen if there are breakdowns?
- Can staff sick leave and holidays become a difficulty?
- Is the experimental schedule unambiguous?
- Have you told other crews about your experiment and are you confident that they would have told you about any experiments that they might be going to conduct? (I have heard of experiments being conducted at the same time by two different groups of people, with the results being of no value because of interference between the two sets of treatments. There is no pedagogical value in knowing who these people are, so they shall remain anonymous.)

7.4 PERFORM RUNS AND RECORD WHAT HAPPENS

Perform experimental runs, repeating runs where results are atypical or unexpected and recording the data tidily and unambiguously.

- Try to avoid departures from intended levels of factors, but record any that occur.
- Record all variables to be measured for each run.
- Do not round off numerical results to units coarser than $\frac{1}{4}$ of the expected error standard deviation.
- Note any calibration of measuring equipment, measurements of the quality of raw material or other information which might prove relevant when a later experiment suggests something as yet unforeseen.
- Check the reasonableness of the data as soon as possible after collecting them – suggestions that there have been minor departures from intended experimental procedures are easier to check the quicker you get onto them.

Some conditions are often to be kept as uniform as possible during the conduct of an experiment. Note these carefully and do what you can to keep them constant. It is generally advisable to complete an experiment quickly so that aspects of the environment and equipment not explicitly considered are kept reasonably constant.

Explain to everyone what is likely to happen and the purpose of the experiment. Encourage recording of information about changes to the process, whether expected or unexpected. Data recording sheets should have space for the date and comments. Emphasize the importance of randomization where the order of runs is being randomized.

Mistakes will still occur. They can be minimized by explaining things carefully as well as providing written instructions. The people conducting the trials should know who to contact if an unexpected result occurs (and know how to decide whether a result is unexpected). They should provide feedback if the experimental plans or protocols do not seem sensible.

Someone who fully understands the experiment and its purpose should visit during the conduct of the runs. The person who will analyse the results (if a different person) should also visit.

People conducting trials should be satisfied that any issues of safety, ethics or waste management have been discussed and will be satisfactorily handled before commencing experimental runs. They should have the right to stop any activities which might endanger themselves or other people.

Statisticians who advise experimenters seldom do any experimentation themselves. Several years ago, as part of a study of mineral sampling procedures, I wanted some data on the bounciness of particles of coal, iron ore, bauxite, limestone, coke and copper ore. With two helpers, I spent about two hours dropping hundreds of particles onto a steel plate and estimating the sizes of their bounces when dropped from a height of one metre. Not very many people saw us, but many of those who did commented that this was the first time they had ever seen statisticians actually doing an experiment.

Example: Long-term breeding trial

When is it OK to abandon an experiment half-way through? Sue Chambers, who was working at the Animal Genetics and Breeding Unit in Armidale, New South Wales related this story during a meeting.

In order to estimate the heritability of a trait in a variety of beef cattle, a long-term breeding trial had been set up. One group of animals was selected to have low values for the trait. After each generation of calves were born, the animals with low values for the trait were used as parents for the next generation. These animals were called the 'low line'. Another group of animals were selected for high values of the trait. These were called the 'high line'.

It was intended to follow this procedure for six generations. The difference in average values of the trait between the high line and the low line is compared to the usual variation in the trait in order to estimate the heritability.

In order to reduce the generation interval in such experiments, it is common practice to put an infertile bull in with the young heifers. (This might be called 'sex training', in contrast to the 'sex education' which my children officially encounter at school.) This bull is called a 'teaser'.

Around the middle of an experiment intended to last for eleven years it was observed that the pregnancies of several heifers were further advanced than expected. Blood tests indicated that they were pregnant to the teaser.

What should be done? It would be possible to analyse the data at the end of the trial in a more complicated way than originally intended, but it was felt that people would have lingering doubts about any such analysis. The hard decision was made to abort all calves sired by the teaser and to extend the duration of the trial by a year.

8

Update Beliefs and Uncertainties

Once you have the data from a set of experimental runs, you need to consider how your beliefs and uncertainties are affected by that data.

An approach to updating beliefs and uncertainties is behind the structure of most of this chapter.

1. You should first cross-examine your data. There may be isolated mistakes and errors. It may be the case that little or none of the data is of value, because the experimental protocol was inappropriate, factor levels turned out not to be the most interesting ones, or there was an earthquake.
2. Provided that the data survives cross-examination, you should focus on the purpose of your experimental programme before choosing between the various formal statistical data analysis techniques.
3. Now think about describing trends. This may be using just the data from your most recent trial or it may be using other data in addition.
4. Then think about describing the variability of the data.
5. Next check the assumptions behind the formal statistical data analysis techniques.
6. Finally, check the statistical significance of any important conclusions which have been drawn.

The final two sections in this chapter deal with the use of computers for statistical analysis and how to decide when to seek professional assistance.

Sometimes experimental data does nothing to reduce your uncertainty, or even makes you feel more uncertain. You may be unsure as to whether your intended experimental protocol is adequate for measuring the thing of primary interest. Perhaps you are not sure that the intended experimental protocol has been followed. Maybe newly acquired data casts doubt about the usefulness of old data.

Sometimes experimental data is of some value, but gives a negative message. It might suggest that the conditions that you have investigated are not promising. It might even suggest that it is time to abandon a particular line of investigation.

Sometimes experimental data is of substantial value. Hopefully, this will happen often. In this case, it is generally appropriate to use a statistical data analysis technique to formally analyse the data. There are many books and computer programs to help you with statistical data analysis techniques. They will help you to do the calculations required to analyse your data and to interpret that statistical analysis.

I do not wish to compete with those books and computer programs. My purpose is to suggest a way of approaching data analysis and to highlight some issues and problems that are associated with analysing data.

In practice, data analysis is often a rather messy process. For instance, it is common for checks of the assumptions behind a formal analysis to suggest that some particular data points be regarded as suspect. It may also happen that, after thinking that the analysis of a set of data is complete, we notice that something is wrong with part of the data and complete re-analysis is necessary. The frustration of re-analysing data can be made less common by carefully checking data quality as the first step in data analysis.

It is common to try several different methods of formal statistical analysis or to fit several different models using a single data analysis method. Deciding when to stop trying alternative models is generally an important issue when this is done.

Sometimes I have heard people say things like 'The formal statistical analysis doesn't highlight anything as significant, but I have a gut feeling that ...' When confronted by such comments, try to interpret them as constructive, suggesting that the model being fitted could in some way be better aligned with the person's 'gut feeling'.

8.1 CROSS-EXAMINING THE DATA

This section discusses ways of checking that the data is not misleading.

Rao (1997, page 70) has suggested that cross-examining data should be a statistician's first task. He argues that this should include reviewing the processes by which data were obtained in order to check for deliberate faking of data, gross recording errors, selection of data or sampling irregularities. These issues should not often be important for experimental data. Nevertheless, a doubting attitude is appropriate.

For experimental data, cross-examining the data includes reviewing the experimental protocols. Sometimes irregularities are not immediately apparent. It is possible, for instance, that while cross-examining one set of data it becomes apparent that a previous set of data should be regarded as suspect.

8.1.1 Checking the Reasonableness of Data

Checks for reasonableness are almost as important as formal analysis. The first things that should be checked for observational data are the cautions given from page 41.

The reasonableness of data should be checked as quickly as possible. This is often done graphically or by simple computational checks. The sorts of checks which might be conducted include the following. Some of them are more useful for observational data than for experimental data.

- Consider the possibility that the data is faked. For instance, Fisher (1936) showed that several of Mendel's experiments on inheritance appear to have been falsified, possibly 'by some assistant who knew too well what was expected', in that the results agree unreasonably closely with theoretical expectations. There are many other instances where it is considered probable that data has been slightly censored, massaged or completely faked in order to support a theory favoured by the experimenter.

UPDATE BELIEFS AND UNCERTAINTIES

- Check that all recorded data is within the bounds of reasonableness. It can be a lot of work to do this thoroughly. However, it is generally a good investment of effort.
- Check that identifications recorded are from the set of identifiers which were to be used.
- Check that different measurements are logically consistent where there are obvious checks. For instance, check that components of exhaustive chemical analyses add up to approximately 100%.
- Use univariate graphical displays or calculate simple descriptive statistics.
- Look for outliers and influential points.
- Look at simple graphical displays. For instance, look at histograms of important variables or combinations of variables, e.g. height, weight, and weight/(predictor of weight given height).
- Consider whether all data points are equally reliable, might have correlated errors, or might have similar biases.
- Distinguish between raw and computed data, and do calculations directly from the raw data whenever possible.

One topic which deserves special attention is missing data. If data is missing, ask why. In some circumstances it may be more useful to know that a data point is missing than it would have been to know the number had it not been missing.

Only if the mechanisms by which data points have gone missing are thought to be unrelated to the value of the response variable (probably due to errors of experimental technique) should the remaining data be analysed as if there had never been any intention to collect the missing data points.

Be sure that you know how unintended missing values, intended missing values, values rejected by consistency checks, and other sorts of missing values are recorded or reported.

Example: Estimating genetic merit of dairy cattle

Before analysing some data on milk yields and parentage of dairy cattle for the purpose of estimating additive genetic merits, a number of checks for reasonableness were done. We rejected data giving zero milk yield or unbelievably high milk yield. We checked for consistency of date of birth, data of start of lactation and date of end of lactation. We checked that parents were born before their offspring. We checked for consistency of breed codes of supposedly purebred cows with the breed codes of their sire and dam, where these were recorded.

We rejected the data on more than half of the records provided to us by the herd-recording systems in the various states of Australia, mostly the historical rather than the current records. However, after the first major data analysis exercise of this type, we found that we had not been thorough enough in our checking. We had treated a large number of cows with unknown sires as if they all had *the same* unknown sire. This was because a computer record which was like that for a real bull, namely '9999: Unknown Friesian' was listed on one of our files of data about bulls.

This problem encouraged us to be more imaginative in thinking about errors that might be present on our data files.

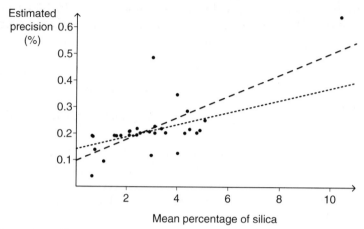

Figure 1. Plot of a measure of sampling and testing precision against the average percentage of silica as reported by various countries. The dashed line was fitted to all of the data. The dotted line was fitted ignoring one data point

- We found eleven supposedly purebred bulls whose breed codes were not consistent with the breed codes of their parents.
- We found a bull which was its own sire, according to our records. (A farmer had re-used the name of a bull in an artificial insemination programme for a son of that bull.)

Example: Cross-examining data in a precision study

Figure 1 shows some information reported as part of a study of sampling and testing precision. The characteristic being considered here is the percentage of silica in iron ores, including pellets. The horizontal axis shows the average grade and the vertical axis shows estimates of a measure of precision including variation due to sampling, sample preparation and measurement.

The dashed line shows the line of best fit to all of the data. The dotted line shows the line of best fit to the data apart from the single point with silica grade above 10%.

Looking at the graph, it can be seen that the point with silica grade above 10% has a great deal on influence on the fitted line. It is also noticeable that there are many points such that the measure of precision is approximately 0.2%. One point with average grade approximately 3% and precision approximately 0.5% might be regarded as an outlier.

However, looking at the data in this way is not 'cross-examining' the data, it is merely looking for possibly suspicious data. Cross-examining means taking a suspicious attitude to all of the data, just as a barrister would cross-examine a witness in a court of law.

What might be wrong with the information which has been plotted as if it were good data? Here are some of the points that seem relevant to me.

- One instruction given to the people providing data was to follow International Standard ISO 3082. Such an instruction is ambiguous because there are three alternative procedures which are considered acceptable by ISO 3082. Furthermore, that Standard was being revised at the time of the data collection, so different versions may have been used.
- Many laboratories which claim to follow Standards do not actually follow those Standards completely, because of poor communication within the laboratories, untrained staff and inadequate equipment. Because of this, the data provides a picture of the precision actually achieved in practice, but does not provide a picture of the precision that would typically be achieved by laboratories following ISO 3082.
- Fourteen data points on Figure 1 are for single shipments unloaded in Japan. The sampling component of variation was estimated using ten pairs of samples for each shipment, but the sample preparation and measurement components of variation were assigned conservative values using Japanese standards. (A Japanese delegate to a meeting of International Standards Organization Technical Committee TC 102– Iron Ores indicated which data was provided by Japan, but other data points were not similarly identified.)
- Only one point is for shipments loaded in Australia. I communicated this data point and know that it summarizes the grade and precision of a large number of shipments. In general, there is likely to be different amounts of data behind different plotted points.
- It is expected that precision will vary with the type of ore being sampled and with the size of cargo being sampled, but this variation is likely to be difficult to distinguish from the country-to-country and laboratory-to-laboratory variation in precision.
- Similar information is available on other aspects of grade. The laboratory which provided the data point with mean silica grade around 3% and precision around 0.5% showed poor precision for other aspects of grade also.

I believe that the data shown in Figure 1 is not adequate to allow reliable conclusions to be drawn about the precision achieved in practice. A useful next step would be to consider qualitative descriptions of the methods used in the laboratories that contributed data. However, this may not be politically possible because of national and company sensitivities.

8.2 FOCUS AGAIN ON YOUR PURPOSE

Before getting lost in the detail of data analysis, stop and think.

Before choosing techniques for formal statistical analysis, think again about your purpose. Keep the real problem and the potential for change in mind.

Example: Visibility of digits

This example is intended to make the point that you should not rush into performing complex statistical analyses without thinking.

Numerous experiments have been done attempting to decide on the most legible types of letters and digits to use in traffic systems; for instance on speed limit signs, number plates and route signs. A report on one such experiment was given by Hind, Tritt and Hoffman (1976). This experiment was a complete seven-factor factorial experiment involving:

- 21 subjects;
- 10 digits: 0, 1, 2, 3, 4, 5, 6, 7, 8 and 9;
- 6 fonts selected on the basis of literature, road practice and earlier research;
- 2 contrasts: namely black digits on a white background or white digits on a black background;
- 5 stroke width to height ratios (SW/H): namely 0.063, 0.083, 0.1, 0.125 and 0.167;
- 5 levels of luminance: namely 1.0, 5.7, 34, 206 and 1233 cd/m^2, and;
- 4 visual angles: 2.31, 1.74, 1.39 and 1.16 milli-radians for 12.7mm high numerals viewed from distances of 5.49, 7.32, 9.14 and 10.97 m.

There were 252 000 data points in all. In analysing this data, it reduced the confusion to realize that subject and visual angle are indicative factors: any sign will be viewed by a variety of subjects from various distances.

Similarly, digit and level of luminance can be regarded as indicative factors. It is not practical to use only the digits which are easiest to read or to standardize the level of luminance. Hence we can concentrate on the average visibility over these four factors.

It is a relatively simple task to look at the trends over the remaining three factors. The fonts differed little. For black digits on a white background, wide digits are best; but for white digits on a black background, narrow digits are best at high luminance, and average width digits are best at low luminance.

8.3 DESCRIBING TRENDS

The part of formal statistical analysis which should be interpreted first is the description of trends. Often a computer package will do other things at the same time as describing trends, but describing trends is usually the first step in interpretation, and is often the first step in deciding what statistical analysis to perform.

8.3.1 Formal Statistical Analysis

The formal statistical analysis of data has at its core the fitting of various models that might be useful. These models may be described as 'statistical' rather than 'mathematical' models, the distinguishing characteristic being that they are concerned to describe possible variation in a response variable, y.

Statistical models are easiest to describe in two parts. For instance, a simple model might be written as $y = f(X, \theta) + \varepsilon$.

The first part of this is a mechanistic model $y = f(X, \theta)$ which gives the expectation or a typical value of the response variable, y, in terms of some observable explanatory variables, X, and some parameters, θ, which will need to be estimated. This part is also sometimes referred to as the trend or as the deterministic or systematic part of the model.

Sometimes X or the variables constituting it are referred to as 'predictor variables', 'exogenous variables' or 'independent' variables. Frequently, the predictor variables can be controlled, in which case they may be referred to as 'factors'. The constants, θ, are often referred to as parameters. The response variable, y, is sometimes referred to as an 'endogenous' or 'dependent' variable.

Luckily, alternative terminology seldom causes much confusion because all we need to worry about is which quantity we are trying to predict or explain, and which quantities we intend to use to help explain or predict it. However, I would discourage use of the terms 'independent' and 'dependent' in this context in order to avoid confusion with the concept of independence.

The second part of a statistical model is the part which describes the variation in the possible values of y for given X. It describes the likely departures from the typical or expected response. It is sometimes described as the 'stochastic', 'random' or 'error' part of the model.

For instance, for the analysis of an agricultural fertilizer trial, the deterministic part of the model might tell us that average crop yield varied with the amount of a fertilizer, the amount of rainfall during a part of the growing season and some aspects of the chemical analysis of the soil. The stochastic part of the model might tell us that departures from this average yield have components due to the measurement of yield, the calibration of the equipment used to analyse soil and the actual pattern of rainfall.

Both parts of a statistical model are generally based on assumptions. Some assumptions are about the mathematical form of relationships. Other assumptions are about the form of probability distributions and the independence or dependence of probability distributions.

Both parts of a statistical model may include parameters which can be estimated. For the mathematical model of the expectation or typical value, it is common to have enough parameters so that the possible mathematical functions $f(X, \theta)$ span the range of forms of relationship thought likely. For the stochastic part of the model, parameters may allow the variability to vary according to controllable factors or from one laboratory to another. A simple form commonly assumed for the stochastic part of the model which has only one parameter is that values of $y - f(X, \theta)$ are independent and normally distributed with variance to be estimated.

8.3.2 Types of Terms used in Models for Trends

What terms should you put into a model for a trend? The short answer is that you should fit models that make sense.

Here, I will make a few comments about models for main effects. These will be followed by some more complicated comments about interactions.

- Where the levels of a factor are discrete, unordered points, like varieties of wheat or makes of motor vehicle in alphabetic order, then it is sensible to fit models in which there is a different average response for each level of the factor. If there is some structure to the set of levels then that structure should be reflected in the models fitted.
- Where the levels of a factor are points on a continuous scale, like time, temperature, mass or stirring rate, it is sensible to use the continuous quantity as the basis for describing models for trends.

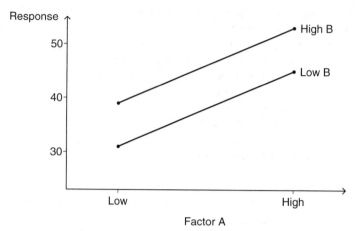

Figure 2. Diagram of responses with no interaction between factors A and B

It often makes sense to use an origin for the continuous variable which is near to the range covered by your data. For instance, a model for the effect of temperature which was being used for temperatures in the range 99°C to 110°C might use 100°C as an origin, so that the model could be interpreted as saying what happens at 100°C and how deviations from 100°C affect the outcome.
- For the intermediate case where the levels of a factor are discrete but ordered, like the number of cylinders in a motor vehicle engine or the number of years schooling that a person has received, then the decision as to whether to use a discrete or a continuous model for trends may be quite difficult to make.

8.3.3 Two-level Factors with no Interactions

Consider the following table of average responses to two factors, A and B, each of which has two levels.

Factor B	Factor A	
	Low	High
High	39	53
Low	31	45

The responses are illustrated in Figure 2.

There is no interaction between the factors. Changing the level of factor A from 'Low' to 'High' increases the response by 14 units, independently of the level of factor B. Changing the level of factor B from 'Low' to 'High' increases the response by 8 units, independently of the level of factor A.

We usually define the term 'main effects' to mean discrepancy from an average of zero. In this case, the main effect of factor A is -7 at the 'Low' level and $+7$ at the 'High' level. The main effect of factor B is -4 at the 'Low' level and $+4$ at the 'High' level.

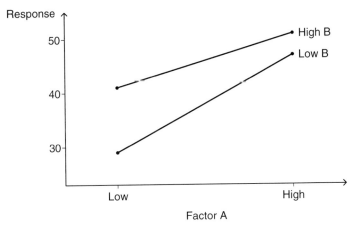

Figure 3. Diagram of responses with interaction between factors A and B

We often write a mathematical expression to express the mean response as a function of the levels of the factors. Denoting the levels of the factors as a and b, with values taken as being -1 and $+1$, this expression is

$$42 + 7a + 4b.$$

A strong suggestion: No matter how you express main effects when giving a written report summarizing experimental results, it is recommended that a graphical report be given as well. This ensures that the magnitude of the effect is not misunderstood.

8.3.4 Two-level Factors with Interactions

Consider the following alternative table of average responses to A and B.

Factor B	Factor A	
	Low	High
High	41	51
Low	29	47

These responses are illustrated in Figure 3.

This time there is an interaction between the factors. Changing the level of factor A from 'Low' to 'High' increases the response by 18 units for the low level of factor B and by 10 units for the high level of factor B. Changing the level of factor B from 'Low' to 'High' increases the response by 12 units for the low level of factor A and by 4 units for the high level of factor A.

When there is an interaction, the usual definition of a 'main effect' is that it is the average effect of the factor, averaged over the conditions investigated in an experiment. Hence we would say that the main effect of factor A is -7 at the 'Low' level and $+7$ at the 'High' level. The main effect of factor B is -4 at the 'Low' level and $+4$ at the 'High' level.

There is an interaction between the factors A and B. It is the discrepancy between the tabulated average responses and the responses which would be expected given the main effects of the factors A and B. It is -2 when A and B are both 'Low', -2 when A and B are both 'High', $+2$ when A is 'Low' and B is 'High', and $+2$ when A is 'High' and B is 'Low'.

We can express the mean response as

$$42 + 7a + 4b - 2ab.$$

The product of the variables a and b can be used as the interaction term. Note that the coefficient of the ab term is negative in this case.

This equation breaks the response up into four terms:

- the average, 42:

Factor B	Factor A	
	Low	High
High	42	42
Low	42	42

- the effect of factor A, $7a$:

Factor B	Factor A	
	Low	High
High	-7	$+7$
Low	-7	$+7$

- the effect of factor B, $4b$:

Factor B	Factor A	
	Low	High
High	$+4$	$+4$
Low	-4	-4

- and the interaction, $-2ab$:

Factor B	Factor A	
	Low	High
High	$+2$	-2
Low	-2	$+2$

8.3.5 Three-level Factors

For three-level factors both main effects and interactions can be more complicated. Consider the following table of average responses.

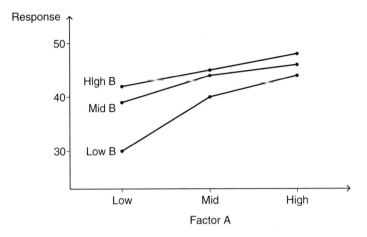

Figure 4. A response affected by factors A and B, each with three levels

Factor B	Factor A		
	Low	Mid	High
High	42	45	48
Mid	39	44	46
Low	30	40	44

The average responses are also shown in Figure 4.

The overall average response is 42. The average influence of factor A can be summarized by saying that the average responses for the levels of factor A are 37, 43 and 46 for the low, mid and high levels, respectively. Comparing these to 42, we say that the main effect of factor A is -5, $+1$ and $+4$ for the low, mid and high levels, respectively.

Similarly, the average responses for the levels of factor B are 38, 43 and 45 for the low, mid and high levels, respectively. As departures from the overall average response, these averages are -4, $+1$ and $+3$.

The interaction between factors A and B tells us how the average response varies with factors A and B, apart from the main effects.

- For this example, the overall average can be expressed in a table as:

Factor B	Factor A		
	Low	Mid	High
High	42	42	42
Mid	42	42	42
Low	42	42	42

- The effect of factor A is:

Factor B	Factor A		
	Low	Mid	High
High	−5	+1	+4
Mid	−5	+1	+4
Low	−5	+1	+4

- The effect of factor B is:

Factor B	Factor A		
	Low	Mid	High
High	−4	−4	−4
Mid	+1	+1	+1
Low	+3	+3	+3

- And the interaction is what is left, namely:

Factor B	Factor A		
	Low	Mid	High
High	+2	−1	−1
Mid	+1	0	−1
Low	−3	+1	+2

Note that the interaction effect always adds to zero in both rows and columns.

Components of effects for 3-level factors

It is sometimes convenient to break main effects for 3-level factors up into linear and quadratic components. These are meaningful if the factors are continuously variable quantities such as time, temperature or the amount of something. They would not be meaningful if the levels were discrete labels such as varieties of plants or types of machines.

For factor A, the main effect was −5, +1 and +4 for the low, mid and high levels, respectively. The linear and quadratic components are often denoted a_l and a_q. The linear component, a_l, takes values −1, 0 and +1. The quadratic component, a_q, takes values +1, −2 and +1. A mathematical expression which summarizes the main effect of factor A is therefore

$$4.5a_l - 0.5a_q.$$

The mathematical expression which summarizes the main effect of factor B is

$$3.5b_l - 0.5b_q.$$

The interaction could be considered to be composed of the four product terms $a_l b_l$, $a_l b_q$, $a_q b_l$ and $a_q b_q$. In this case it is

$$-2a_l b_l + 0.5a_l b_q + 0.5a_q b_l + 0.0a_q b_q.$$

Table 1. Data from a simulation game

Levels of factors		Yields	
A	B	Rep. 1	Rep. 2
40	3	73	71
40	4	117	112
40	5	128	124
40	6	82	85
50	3	100	104
50	4	139	134
50	5	110	110
50	6	57	59
60	3	119	117
60	4	126	131
60	5	93	89
60	6	58	55

In general, the coefficients can be found by fitting a regression model. This is sometimes useful if the linear×linear interaction is more likely to be important than the others, but in such situations response surface designs might be more appropriate than main effects designs.

8.3.6 Extensions

- Missing data or unbalanced designs can be handled by fitting regression models.
- There are differences between categorical and continuous factors which become important when there are three (or more) levels for a factor. With continuous factors, the decomposition into linear and quadratic components is usually helpful. For categorical factors, it is more likely to be useful to decompose a main effect into components such as the following:
  $-1, +1, 0$, the difference between two categories, and
  $-1, 0$ and $+1$, the difference between another pair of categories.
  Alternatively, the second component might be:
  $-1, -1$ and $+1$, the difference between a pair of categories and the third category.

Example: Simulated data

Consider the data in Table 1 from a small factorial experiment (actually generated using a simulation game).

A regression package gave the following fitted regression coefficients and standard errors:

estimate	s.e.	parameter
124.2	3.847	Constant
-0.025	0.221	$A - 50$
-11.15	1.61	$B - 4.5$
-0.0287	0.0382	$(A - 50)^2$
-18.04	1.80	$(B - 4.5)^2$
-1.340	0.197	$(A - 50)(B - 4.5)$

The model being fitted is a quadratic surface with the origin being moved to the centre of the experimental region: $A = 50$, $B = 4.5$. The fitted constant tells us the estimated level of the surface at this origin. It is much easier to interpret than would be a constant when the origin were not shifted, because there is great uncertainty about the yield for $A = 0$, $B = 0$.

The linear terms tell us the estimated slope of the fitted surface at the origin. The second-order terms tells us how the slope of the fitted surface varies. The things to notice are as follows.

- The estimates of the coefficients for $(A - 50)^2$ and $(B - 4.5)^2$ are both negative, giving some hope that the fitted curve will have a well-defined maximum.
- The estimate of the coefficient for $(A - 50)^2$ is smaller in size than its standard error, indicating that we cannot be confident on the basis of this experiment that the coefficient is negative.
- Taking partial derivatives with respect to $A - 50$ and $B - 4.5$, we see that the slope in the A direction is

$$-0.025 - 0.0575(A - 50) - 1.340(B - 4.5)$$

and the slope in the B direction is

$$-11.15 - 1.340(A - 50) - 36.08(B - 4.5).$$

Solving the simultaneous equations in which both of these are set to zero tells us where the stationary point of the fitted surface lies. In the light of the uncertainty in the coefficient for $(A - 50)^2$, this is of little value for the present example.

8.3.7 Multiple Comparisons and Use of Contrasts

Some factors have discrete levels which have no natural ordering. For instance, an experimenter might have no prior knowledge about cultivars of a flowering plant. When analysing an experiment involving such a factor it is appropriate to make use of one of several tests which have been suggested for handling 'multiple comparisons'.

The problem which multiple comparisons tests are designed to overcome is that there are a large number of pair-wise comparisons which might be made, so we must expect that an average of 5% of the differences which are really zero will appear to be statistically significant according to any pair-wise test which has a 5% error rate. For instance, in an experiment involving 15 cultivars there are 105 pair-wise comparisons so an average of $105 \times 0.05 = 5.25$ differences will falsely appear to be statistically significant if a statistical test with a pair-wise error rate of 5% is applied.

There are several different tests for multiple comparisons. They have slightly different properties. If you often use one of them then I would encourage you to continue using the same one so that your intuitive understanding of it remains useful.

Where levels of a treatment factor have a logical structure rather than being merely a set of categories, the use of multiple comparison procedures is strongly discouraged. If analysing such experiments using Analysis of Variance, partition the treatment sums of squares to show how much variation is associated with particular comparisons or contrasts. This answers the right questions. The tests resulting from such partitioning are more likely to give statistically significant and substantively significant results than if partitioning were not used.

Ask yourself whether your treatments are like a number of varieties of wheat in that there is no prior order or relationship between them. If the answer is 'Yes', then a basic ANOVA will do. If the answer is 'No', then you should extract contrasts. For instance, in analysing data from an interlaboratory trial with eight laboratories, five of which are very familiar with a procedure and three of which seldom use that procedure, it might be interesting to know how consistent the results were for the five laboratories familiar with the procedure. Agreement between the results from those five laboratories and the results from the other three laboratories may be less important, or at least lead to different classes of possible action.

Bryan-Jones and Finney (1983) summarizes these issues very well. There is an editor's note at the end of that paper saying that the statistical guidelines for the Journal of the American Society for Horticultural Science and for HortScience have been changed to recommend the use of multiple comparisons 'when treatments consist of a set of unrelated materials such as cultivars or chemicals'. I also recommend the allegorical story about Baby Bear's experiment in Carmer and Walker (1982).

8.3.8 Non-linear Models for the Typical Value

When fitting non-linear models, most of the disasters that I have ever seen could have been avoided by following the following two suggestions.

- Use models that make sense. This means considering the relevant science and technology, not just looking at the data. Also check that the models seem reasonable intuitively. If it seems relevant, also check that the derivatives of the models are intuitively sensible.
- Use mathematical forms for models which make individual parameters easy to interpret, rather than using mathematical forms which look simple in that they are mathematically tidy.

One type of model which is commonly fitted but which does not make sense is to fit an effect for the amount of some class of substance added to an effect which is different for each substance in that class. Such models imply that adding none of one substance will have a different effect from adding none of a different substance! For instance, in Taguchi's analysis of the Ina tile experiment (see page 181), the effect of adding no feldspar when Mikumo feldspar would have been used is different from the effect of adding no feldspar when Mihanayama feldspar would have been used.

There are two other silly models that I have seen used for analysis of agricultural fertilizer trials. One silly model is to fit a term in $\sqrt{x}$ to the mean yield per hectare where x is the amount of some fertilizer. Such a model implies that the marginal effect of fertilizer, which is the derivative of that model, includes a term in $1/\sqrt{x}$. This term becomes arbitrarily large as x tends to zero, ensuring that the economically optimum amount of fertilizer will be greater than zero.

Another silly model is to fit a term which is zero if no fertilizer is used and which is one if any fertilizer is used. If this term has a positive coefficient then the model implies that using, say, one microgram per hectare of the fertilizer would have a worthwhile effect.

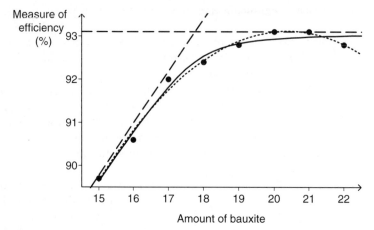

Figure 5. Plot of data for an experiment on efficiency of dissolution of bauxite. A quadratic curve of best fit is shown as a dotted line. A non linear curve of best fit is shown as a solid line, with its asymptotes shown as dashed lines

Both of the above issues are illustrated by the following example.

Example: Dissolution of bauxite

Consider the following hypothetical experiment on dissolution of bauxite, which has some of the features of a real but confidential example. Starting with 15, 16, 17, 18, 19, 20, 21 and 22 g of bauxite gave the percentage process efficiencies of 89.7, 90.6, 92.0, 92.4, 92.8, 93.1, 93.1 and 92.8, respectively. These are plotted in Figure 5.

The quadratic model with the mathematically tidy equation

$$y = a + bx + cx^2$$

which best fits the data (in the sense that the sum of squared residuals is a minimum) has estimated coefficients $\hat{a} = 43.40$ (with standard error of 4.12), $\hat{b} = 4.88$ (s.e. 0.45) and $\hat{c} = -0.120$ (s.e. 0.012). The quadratic model of best fit is shown as a dotted curve on Figure 5.

The value $\hat{c}$ is straightforward to interpret as twice the second derivative of the curve. The value $\hat{b}$ can be interpreted as the first derivative of the curve at $x = 0$, or when no bauxite is added. Similarly, $\hat{a}$ can be interpreted as the value of y at $x = 0$. The last two parameters do not seem very relevant to interpreting the model in the region of interest, which is about the same as the region for which data has been obtained. Also, their standard errors do not seem relevant to the region of interest.

An additional symptom which indicates that this model is likely to be difficult to interpret is that the correlations of the parameter estimates are near to unity in absolute value. The correlations are -0.9984 for $\hat{a}$ and $\hat{b}$, 0.9939 for $\hat{a}$ and $\hat{c}$, and -0.9985 for $\hat{b}$ and $\hat{c}$.

A mathematical equation which has the same scope for fitting quadratic curves is

$$y = a + c(x - b)^2.$$

UPDATE BELIEFS AND UNCERTAINTIES

Fitting this model gives estimated parameters $\hat{a} = 93.12$ (with standard error of 0.08), $\hat{b} = 20.39$ (s.e. 0.22) and $\hat{c} = -0.120$ (s.e. 0.012). The value $\hat{a}$ can be interpreted as the maximum value of y, and $\hat{b}$ can be interpreted as the location of this maximum.

The standard errors of $\hat{a}$ and $\hat{b}$ are of obvious relevance and interest. The standard error of $\hat{a}$ describes how confident we should be about the maximum value for y, and the standard error of $\hat{b}$ describes how confident we should be about the location of this maximum.

From previous practical experience and research on this topic, a quadratic curve was not thought likely to be a good fit, other than in a small neighbourhood of the maximum. It was expected that the efficiency measure would be approximately constant for large amounts of bauxite and would be approximately linear for small amounts of bauxite. Making use of this background information, one attempt to fit a non linear model fitted the hyperbola shown as a curved solid line. Its asymptotes are the straight dashed lines and have equations $y_1 = c$ and $y_2 = a + bx$. Given y_1 and y_2, the value for the hyperbola depends also on a distance parameter, d, and is $y = \frac{1}{2}(y_1 + y_2 - \sqrt{(y_1 - y_2)^2 + d^2})$.

Parameter estimates were $\hat{c} = 93.1$ (s.e. 0.25), $\hat{a} = 62.3$ (s.e. 4.7), $\hat{b} = 1.205$ (s.e. 0.308) and $\hat{d} = 1.39$ (s.e. 1.32).

An alternative mathematical formulation of this non linear model was to replace the equation $y_2 = a + bx$ for one of the asymptotes with the equation $y_2 = c + (1+b)(x-a)$. This is still a linear function of x, but the parameters have interpretations which are of more direct interest.

- The parameter b now describes the difference between the slope of the asymptote and unity. This is of interest because there were differences of opinion amongst the experts as to whether that asymptote was necessarily of unit slope. The parameter estimate was $\hat{b} = 0.205$ (s.e. 0.308) and the difference from zero is not statistically significant.
- The parameter a now describes the location of the point of intersection of the asymptotes. This is estimated to be at $\hat{a} = 17.76$ (s.e. 0.67).

A point to note in comparing the quadratic and hyperbolic models is that the residual standard error is in fact smaller (0.157) for the quadratic model than for the hyperbolic model (0.209). This reason for possibly preferring the quadratic model should be dismissed as being of lesser importance than the fitting of a model that makes sense.

8.4 DESCRIBING VARIABILITY

The second part of interpreting a formal statistical analysis is to look at how the variability or the noise is described. There may be several components to this description. For experiments which are investigating sampling and measurement errors, the mathematical model for the trend may be a constant, so the description of the variability appears to be the first step.

The simplest model which is used to describe variability is that the departures from the model for the trend are independent and have constant variance. In this case, all you need to do to describe the variability is to estimate the variance of the noise.

Alternatively and equivalently, you can estimate the standard deviation of the noise. This is equivalent because the variance is the square of the standard deviation. It is a very practical alternative because the standard deviation is on the same scale as the response variable. It therefore has the same units as the response variable and is easy to understand.

Example: Simulated data (continued)

Continuing the example using the simulated data shown in Table 1 on page 211: An estimate of the standard deviation of the residuals is 8.82 units of yield.

An analysis of variance table might be part of the formal statistical analysis. It would show that the residual sum of squares was 1401 on 18 degrees of freedom, giving a residual mean square of 77.83. Taking the square root of the residual mean square, we see that the estimate of the standard deviation of the noise associated with individual measurements is 8.82.

Monitoring variability

When a series of experiments are conducted, it is good practice to monitor the estimates of residual standard deviation. These can be expected to be reasonably stable if the likely causes of variation (variations in weighing, measuring, calibrating, sampling, setting up runs, and so on) are likely to have stable variance and the experimental outcomes are equally sensitive to them. If an experiment is observed to have a residual standard deviation which is unusually high or low then it is worth asking whether something has been done particularly poorly or particularly well.

8.4.1 Analysis of Experiments for Estimating Components of Variance

When analysing experiments in which there are two or more components of variance it is crucial to focus on the problem of estimating the amount of variability for those components. Do not focus on the interpretation of any particular procedure for analysing data.

Example: Testing bitumen

Table 2 presents some data from Dickinson and Robinson (1977). (The data are for Material A, no needle treatment.) For this data there are 12 laboratories. These correspond to columns in Table 2. In each laboratory, two operators each tested two samples. Results for operator 1 are shown in the top half of Table 2; results for operator 2 are shown in the bottom half. Each sample was tested three times. Within each column, results are given for the three tests on one sample followed by the three tests for the other sample.

Figures 6, 7, 8 and 9 illustrate the type of graphs which might form part of a graphical approach to the consideration of this data. Note that all four graphs have similar scales on the vertical axis.

Figure 6 shows all of the data plotted against the laboratory identifier. Because penetrations are only measured to the nearest tenth of a millimetre and this is not

UPDATE BELIEFS AND UNCERTAINTIES

Table 2. Data from interlaboratory trial of penetration test for paving bitumens

1	2	3	4	5	6	7	8	9	10	11	12
18.7	18.2	18.2	17.3	18.4	18.3	17.8	17.7	17.9	17.2	17.5	18.1
18.7	18.3	18.6	18.3	18.1	18.2	17.9	17.6	17.9	17.4	17.6	18.3
18.8	18.4	18.3	19.4	18.2	18.1	18.0	18.2	18.0	17.2	17.8	18.5
18.4	18.2	18.7	17.5	18.1	18.4	18.3	17.6	18.0	17.3	17.7	18.1
18.6	18.2	18.4	18.0	17.9	18.3	18.7	17.1	17.8	17.1	17.9	18.5
18.8	18.4	18.7	18.3	17.7	18.1	18.7	17.8	17.8	17.3	17.9	18.7
18.0	18.8	17.9	17.2	18.0	18.0	18.1	17.9	17.6	17.4	17.6	18.4
18.1	18.6	18.2	17.6	18.1	18.2	18.5	17.9	18.2	17.1	17.9	18.5
18.4	18.5	18.3	18.4	18.1	18.4	18.7	17.9	17.7	17.3	17.7	18.8
18.0	18.4	17.7	18.0	17.8	18.1	18.2	18.0	17.4	17.3	17.8	18.2
18.1	18.2	18.4	18.1	18.3	18.2	18.6	17.9	18.2	17.4	17.9	18.8
18.1	18.4	18.7	19.1	18.2	18.3	18.8	18.2	17.5	17.2	17.7	18.8

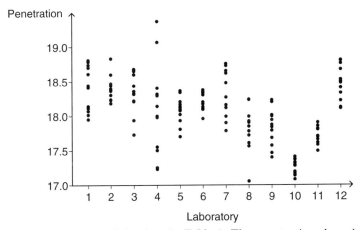

Figure 6. Plot of all of the data in Table 2. The penetrations have been dithered so that data points are unlikely to hide one another

trivial compared to the variation between results, it is common for several results in a laboratory to be identical. Results have been dithered so that data points are less likely to be obscured by others.

From Figure 6 we can see that there is a substantial amount of variation between laboratories.

Figure 7 shows the averages over samples and replicate measurements. It allows us to see the amount of variation between laboratories and the amount of variation between operators.

Figure 8 shows the differences between sample 1 and sample 2 for each operator in each laboratory. This gives us some idea of the amount of variation between samples.

Figure 9 shows the differences between replicate 3 and replicate 1 for each operator on each sample in each laboratory. It is clear that the differences for laboratory 4 do

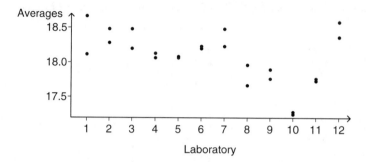

Figure 7. Average penetration measurements for each operator

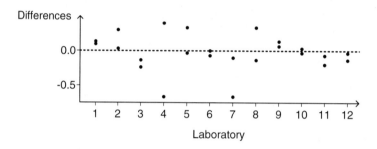

Figure 8. Differences in average penetration between different samples for each operator

not have an average value of zero. Looking at the raw data in Table 2 on page 217 indicates that later replicates gave higher penetration measurements for laboratory 4.

Reminder: This seems an appropriate time for me to remind readers that I have titled this chapter 'Update Beliefs and Uncertainties' rather than, say, 'Analyse your Data'. Having looked at the data, you may suspect that some operators need training or that some laboratories may have inferior equipment or are not strictly following the test method. Such considerations are part of the updating of your beliefs and uncertainties.

8.4.2 Separating Variation and Trends

Sometimes, the two parts of a statistical model may be difficult to separate.

One problem arises when data is censored or selected, because the estimated effect of the censoring depends on estimates of the components of variability. For instance, suppose that we have data on the first and second lactation milk yields for some dairy cows. If some cows have been culled after their first lactation, with the choice of animals to be culled being partly based on first lactation milk yield, then it is not possible to estimate the average difference between second lactation milk yield and first lactation milk yield without simultaneously considering the correlation between first and second lactations.

UPDATE BELIEFS AND UNCERTAINTIES

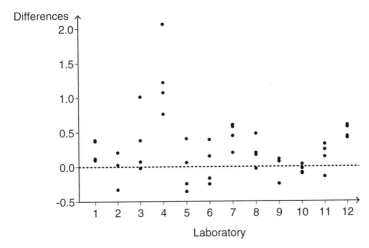

Figure 9. Differences in penetration between replicate 3 and replicate 1 for each operator on each sample in each laboratory

The difference between all first lactations and all second lactations is a biased estimate because the cows culled would have produced less milk on average than the cows which completed their second lactation. The difference between the first lactations and second lactations for cows completing two lactations is also a biased estimate, but the logic for seeing this is more subtle. Cows with equal potential to produce in later lactations are more likely to be culled if their first lactation milk yield is poorer than commensurate with their potential, and are less likely to be culled if their first lactation milk yield is better than commensurate with their potential. Hence, first lactation milk provides an overestimate of potential. See Henderson *et al.* (1959) for more details on this situation.

8.5 CHECKING ASSUMPTIONS

This section discusses how to think about checking the assumptions behind a formal statistical analysis.

Checking the assumptions behind data analysis is a large and complicated topic. It is always important to spend a little time thinking about assumptions. Checking the assumptions behind the formal analysis includes checking that the typical response does not depend on factors or other terms not included in the model; checking for outliers amongst the residuals, that the residual variance is approximately constant, and that residuals are independent.

8.5.1 *Normality*

Checking for normality is not an issue in most circumstances. Normality is often best regarded as an approximation, rather than as an assumption. People often feel a need to check their assumption of normality when what they should be checking is the robustness of their statistical procedures.

Example: Okra variety trial

Some varieties of okra were being compared in a variety trial. A residual plot showed that the distribution of residuals was skewed to the right. This means that a small number of plots produced very much larger yields than average (after allowing for the differences between varieties). Skewness to the left is much more common in plant yield data, with the very best plots being only slightly better than average but some plots having very low yields, due to local problems as disease, water-logging or infertile soil.

Commercially, pods of okra would be harvested as soon as they reached a standard size. The experimental okra had been harvested on Mondays, Wednesdays and Fridays, without caring about the uniformity of sizes of pods.[1] Sometimes pods were allowed to become very large before they were harvested. These had large weights and tended to skew distributions of yields (recorded as the mass of pods harvested) to the right.

The key to the solution to this problem was to consider the economic value of the okra harvested. This was thought to be more closely related to the number of pods harvested than to the mass of pods harvested. Data on the number of pods harvested was available, was used for the formal statistical analysis and was presented as the most important measure of yield, considering the nature of the harvesting regime.

Homoscedasticity: That the error variance does not vary over the range of interest is often one of the most important assumptions. There are standard ways of checking for equality of variances.

8.5.2 Outliers and Influential Points

Graphical or computational checks for outliers and influential points are a good way to test a broad range of assumptions. Such checks are often done first.

Consider a set of artificial examples, four of which are similar to ones in Anscombe (1973). Suppose a regression analysis program produced output indicating the following points:

- number of observations = 15;
- mean of the x values = 8.0;
- mean of the y values = 5.8663;
- equation of regression line is $y = 0.838 + 0.6285x$;
- regression sum of squares = 110.617 (on 1 degree of freedom under null hypothesis);
- residual sum of squares = 39.1 on 13 degrees of freedom;
- estimated standard error of slope of regression line = 0.6285;
- multiple R squared = 0.7388;
- correlation coefficient = 0.8595.

---

[1] Government policy required that money from the sale of produce from agricultural research stations go into consolidated revenue. Because such money was not returned to the individual research stations, they had little incentive to maximize the commercial value of their produce. Consequently, experimental crops were often weeded and harvested with much less care than would be taken by a commercial farmer. This was sometimes a barrier to the acceptance of research results, because farmers were reluctant to believe people who ran their own farms poorly.

UPDATE BELIEFS AND UNCERTAINTIES

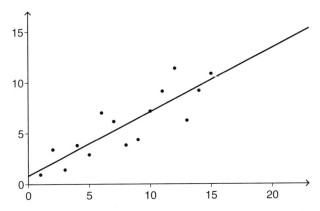

Figure 10. Scatter plot that might be expected given the regression analysis. The solid line is the regression line

Note that these statements could all be deduced starting from six facts. There are 15 observations. The sum of the x values is 120. The sum of the y values is 87.995. The sum of the squares of the x values is 1240. The sum of the squares of the y values is 665.93 The sum of the products of corresponding x values and y values is 879.95.

Comparison of scatterplots for seven data sets which could have produced this regression output will help us to understand the meaning of the numbers produced by regression analysis and the restrictions on the usefulness of the technique. In each case, the regression line is shown as a continuous line.

Figure 10 illustrates the kind of scatter that might be expected when the correlation between two variables is investigated. The theoretical description and the assumptions made are perfectly appropriate.

Figure 11 suggests clearly that there is a quite close relationship between x and y but that this relationship is not linear. The underlying theoretical model that there is a linear model with random errors associated is not reasonable. A quadratic model like that shown by a dotted line would be much better.

Figure 12 illustrates the concept of an outlier. One data point is sufficiently far removed from the general pattern suggested by the others that it seems desirable to give a different explanation for it. Possibly it is simply an error of data collection. If so, then the most appropriate way to treat it is to simply throw it away. Another possibility is that the process which produced the data is unreliable in that it occasionally produces outputs which do not fit the normal pattern provided by the bulk of the items. In such cases, the most useful way to use the data might be to discard the bulk of the data and to investigate the pattern of the outliers.

Figure 13 illustrates the concept of an influential point. One data point provides a large proportion of the information about the position of the line of best fit. In this case, if that one point were removed then the slope of the line of best fit for y as a function of x would be very imprecisely estimated.

Figure 14 illustrates a situation where an outlying point has very little influence on the line of best fit. In this case the point could be described as an outlier but not as an influential point. In contrast, the outlier in Figure 12 is influential.

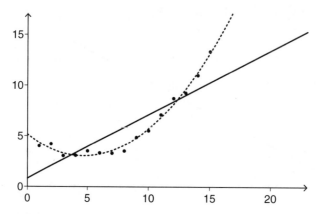

Figure 11. Scatter plot suggestive of a quadratic relationship. The solid line is the regression line. The dotted line shows the best-fitting quadratic model

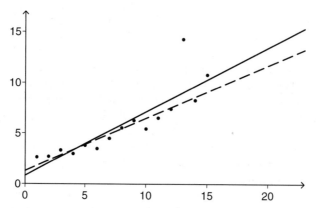

Figure 12. Scatter plot showing an influential outlier. The dashed line shows the regression line if that point is omitted

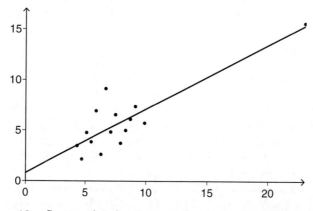

Figure 13. Scatter plot showing an influential point which is not an outlier

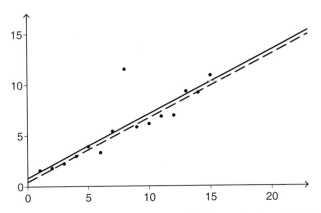

Figure 14. Scatter plot showing an outlier which is not influential. The dashed line shows the regression line if that point is omitted

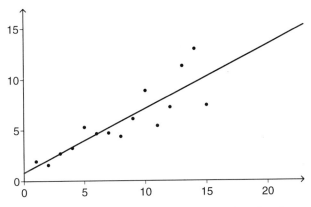

Figure 15. Scatter plot suggestive that the scatter in values for y increases with increasing x

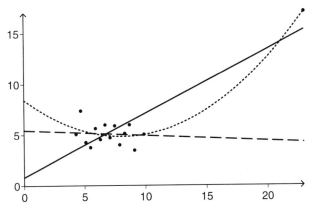

Figure 16. Scatter plot showing an influential point which is outside the range of the other values of the independent variable. A quadratic fitted to all of the data is shown as a dotted line. The regression line omitting the influential point is shown as a dashed line

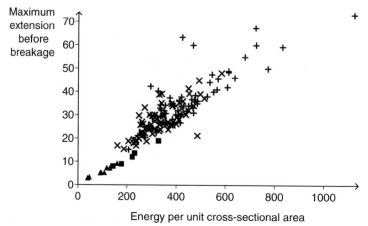

Figure 17. Scatter plot of two measures of strength for coupons of polyethylene pipe which include welds

Figure 15 suggests that the amount of scatter about the regression line may not constant across the range of the data. Perhaps the variation increases with x; perhaps it increases with y.

Figure 16 shows a situation where there is an influential point which needs to be discussed very carefully. The line of best fit to all of the data is shown and has positive gradient. The line of best fit to all of the data except that point would have negative gradient. Perhaps the linear model is too simplistic.

Experimental data would not usually result in a scatter plot like Figure 16, because the design of most experiments guarantees that controllable variables will be approximately balanced. Observational data might result in a scatter plot like Figure 16. For instance, crop yield might generally increase with increasing rainfall except for a single season when unusually heavy rainfall caused devastation or mould.

Example: Tensile testing of welds in plastic pipe

Figure 17 shows a scatter plot which was produced during cross-examination of some tensile testing data. The samples being tested were coupons cut from welded high density polyethylene pipe.

The testing procedure monitors the tensile force as coupons are pulled at a steady rate, which is generally 5 mm/sec. The resulting graphs of force against extension (usually referred to as 'stress–strain graphs') can be used to calculate several different summary statistics. Two of these summary statistics are plotted against one another on Figure 17.

Four different plotting symbols have been used on Figure 17. Addition and multiplication signs have been used to differentiate two different gauges of tensile coupon. It appears that the samples corresponding to addition signs achieve higher tensile testing results, according to both of the summary statistics 'energy per unit cross-sectional area' and 'maximum extension before breakage'. Whether this is due to differences in test methods or in the tensile strength of welds tested in different laboratories is not clear from the graph.

The multiplication sign near (energy = 500, extension = 20) was due to an error in calculations using a spreadsheet. The error affected several other points also.

The points plotted as solid triangles correspond to welds which were considered to have suffered brittle failure and which were tested at the usual speed of 5 mm/sec. For these welds the two summary statistics 'energy per unit cross-sectional area' and 'maximum extension before breakage' are linearly related.

The points plotted as solid squares correspond to six welds which were tested at 50 mm/sec rather than the usual speed of 5 mm/sec. One of these welds did not fail in a brittle manner, it is above a trend line which would fit to the welds which failed in a brittle manner.

A simple example

When looking for outliers, it is important to remember that outliers affect the fitted model. Hence outliers cannot be very outlying for small data sets and are unlikely to be very outlying for complicated models.

Consider the following example which illustrates that extreme outliers cannot happen for small data sets when non robust procedures are used for model fitting.

The data consists of nine replicates of '-1' and one result which is '$+9$'. The mean is '0', so the residuals are the same numbers as the data. The estimated residual standard deviation is equal to the sample standard deviation which is $\sqrt{10} = 3.16$, so the residual of '$+9$' is less than three times the residual standard deviation.

The influence of outliers on saturated designs

In the set of examples used in the earlier section about regression, we were fitting a model with two parameters to 15 data points. When fitting models which have large numbers of parameters compared to the number of data points it can be very difficult to detect outliers.

To illustrate the difficulty of detecting outliers when fitting highly saturated models, meaning models in which the number of parameters is almost as large as the number of data points, models were fitted to a data set of 16 points in which all points were zero except the last, which was -16. The experimental designs were fractional factorials. The numbers of parameters to be fitted are each one more than the number of factors in the design, since an overall mean parameter is also fitted.

Table 3 gives the numbers of parameters fitted for these various models and the residuals obtained. The data had only one outlier, the last data point, but fitting of a model means that residuals at other data points are also affected. For this example, we need to have twice as many data points as parameters before the real location of the outlier is clear in that the residual at the location of the real outlier is twice as large as the other residuals.

Example: Mode choice after Tasman bridge accident

This is an illustration of applying the ideas of outliers and influential points in a non standard way.

Table 3. Table of residuals for models in which there is only one outlier. The last data point is at −16, but the last residual is smaller than 16 in magnitude when many parameters are being fitted. Also, the location of the aberrant point is not obvious because there is no uniquely largest residual when 12–15 parameters are being fitted

No. par.	Residuals corresponding to the 16 data points															
15	+1	+1	+1	+1	+1	+1	+1	+1	−1	−1	−1	−1	−1	−1	−1	−1
14	+2	+2	+2	+2	0	0	0	0	0	0	0	0	−2	−2	−2	−2
13	+3	+3	+1	+1	+1	+1	−1	−1	+1	+1	−1	−1	−1	−1	−3	−3
12	+4	+2	+2	0	+2	0	0	−2	+2	0	0	−2	0	−2	−2	−4
11	+3	+3	+3	−1	+3	−1	−1	−1	+3	−1	−1	−1	−1	−1	−1	−5
10	+4	+2	+2	0	+2	0	0	−2	+4	−2	−2	0	−2	0	0	−6
9	+5	+1	+1	+1	+3	−1	−1	−1	+3	−1	−1	−1	−3	+1	+1	−7
8	+4	+2	+2	0	+2	0	0	−2	+2	0	0	−2	−4	+2	+2	−8
7	+5	+1	+3	−1	+1	+1	−1	−1	+1	+1	−1	−1	−3	+1	+3	−9
6	+4	+2	+2	0	+2	0	0	−2	0	+2	−2	0	−2	0	+4	−10
5	+3	+3	+1	+1	+1	+1	−1	−1	+1	+1	−1	−1	−1	−1	+5	−11
4	+4	+4	0	0	0	0	0	0	0	0	0	0	0	0	+4	−12
3	+3	+3	+1	+1	+1	+1	−1	−1	−1	−1	+1	+1	+1	+1	+3	−13
2	+2	+2	+2	+2	0	0	0	0	0	0	0	0	+2	+2	+2	−14
1	+1	+1	+1	+1	+1	+1	+1	+1	+1	+1	+1	+1	+1	+1	+1	−15

Having been hit by a ship, a bridge across the Derwent River near Hobart could not be used for a substantial period. The choices made by people who had previously used this bridge for their journey to work were studied and models were fitted to them.

Four modes of travel were considered, classified according to the mode of transport in use at the point of crossing the river.

- Driving alone. The length of the journey to work was substantially longer than before because a bridge further upstream had to be used.
- Driving with at least one other person, whether always in the same vehicle or whether vehicles were alternated.
- Taking a ferry across the river.
- Travelling by train. (The train crossed the river on a bridge near to but different from the one which had been hit.)

There was some concern that the models being fitted using data from about 3000 people might be being strongly influenced by a small number of observations. The models being fitted were multi-dimensional logit models in which the probability of choosing a particular mode of transport depended on several variables such as estimated time of travel from home to a railway station and estimated time to walk from the city ferry terminal to the place of employment.

Rather than thinking about the distance between an observation and a fitted model (which can never exceed unity because the observations can only be zero and one), we thought about the probability of the observed mode choice according to the fitted model. For most people in the data set, the fitted model gave the probability of their observed mode choice as a moderate number, such as 0.2, 0.5 or 0.9.

However there was one person whose mode choice was a very poor fit to the model. He (or she) appeared to choose to drive with someone else despite the fact that this took about 100 minutes longer than driving alone! The modelled probability of the observed choice was only about 0.01, even though this piece was data had been used in the fitting of the model. Within the small group of people involved in fitting these models, our favourite hypothesis was that he (not she) was driving by an indirect route in order to be accompanied by his mistress.

It turned out that there had been poor interfacing between two stages of data preparation. In one place, '1 5' had been interpreted as 105. Another piece of computer code scanned from right to left until it found a blank and interpreted '1 5' as 5. While this was not as titillating as our initial hypothesis, this explanation of the data which appeared to be a poor fit to the model was of substantial benefit to the project.

Residuals and influential points in hierarchical trials

In split-plot designs and in hierarchical designs for estimating components of variance it is possible to have residuals and influential 'points' at different levels of the hierarchical structure, which are sometimes referred to as 'strata'.

For instance, consider an interlaboratory trial investigating the precision of a test method. It might be found that one laboratory or a single operator at one laboratory has obtained results that are quite different from all of the others. The set of results for that laboratory or operator might be regarded as not being typical of the precision to be expected for the test.

For an industrial experiment in which some factors involve the setting up of equipment and other factors can be varied within a set-up, it is possible that all of the runs for a single set-up might be regarded as outlying or as influential.

8.5.3 Testing Independence

A distributional assumption which is often violated by enough to matter is the assumption that errors are independent. This assumption can be tested by plotting the residuals from a model in sequence and looking for trends. Formal tests for autocorrelation can also be used.

The assumption of the independence of errors can also be considered by thinking about the processes by which errors of sampling, measurement, calibration and other sources of variation are likely to arise. Would you expect that some sources of variation will affect groups of results? Might there be likely to be serial correlation because of trends in soil fertility or in characteristics of raw materials? In my experience, thinking about the processes involved in getting data is more useful than merely looking at the data.

Simulation using various structures for generating noise can be useful for investigating the sensitivity of a data analysis procedure to the assumption of the independence of errors.

It is possible to analyse experimental data without making the assumption that errors are independent. However, I would advise users of such procedures to be wary

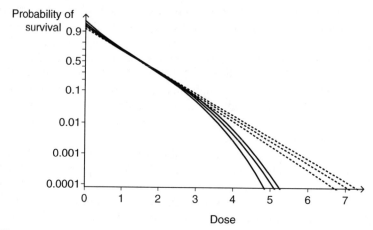

Figure 18. Predicted values and confidence bounds for logit and probit models. Dotted lines are for the logit model. Solid lines are for the probit model

that such analysis procedures may tend to overstate confidence levels because they do not consider uncertainty in the estimates of the parameters which describe the distribution and autocorrelation structure of the errors.

8.5.4 What do you do when the Assumptions are not Satisfied?

If you discover that some of the assumptions behind statistical procedures are not satisfied then you have a few options.

- You may use non parametric data analysis procedures which are, generally speaking, less sensitive to the effects of interest as well as being less sensitive to distributional assumptions.
- You may be able to remove data points which are outliers or make a transformation and then analyse the data again. Much data analysis in practice is repeated in this way.
- You may wish to check the robustness of your chosen data analysis technique to the assumptions which are violated. This can often be done by simulating situations which have similar properties to your data and applying the data analysis technique to each simulation.
- You may express the sensitivity to an assumption as an additional component of uncertainty.

Example: Comparing logit and probit models

A large number of individual plants were exposed to various doses of a chemical. Of 5000 plants exposed to a unit dose, 1500 died. Of 5000 plants exposed to twice the unit dose, 3500 died. Of 100 plants exposed to seven times the unit dose, all 100 died. These results were analysed by fitting generalized linear models with both logit and probit link functions.

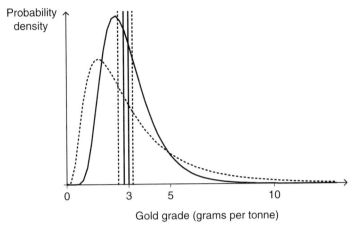

Figure 19. Two possible distributions for gold grades. The one shown as a continuous line is log-normally distributed with standard deviation of 0.4 on the logarithmic scale, mean 3.00 and median 2.77. The one shown as a dotted line is log-normally distributed with standard deviation of 0.7 on the logarithmic scale, mean 3.20 and median 2.50. The means and medians are shown as vertical lines

Figure 18 shows some predictions from the fitted models. The dotted lines show the fitted values for the logit model, together with confidence bounds at plus and minus twice the standard error of the fitted values. The continuous lines show the fitted values for the probit model, together with confidence bounds at plus and minus twice the standard error of its fitted values.

It is easy to see that for doses greater than about 3.5 times the standard dose the confidence regions for the logit and the probit models do not overlap. This illustrates the point that when communicating our uncertainty about the predictions it is important to mention uncertainty about the form of the model.

The region where the two models give substantially different predictions is outside the range of most of the data. In fact, the 100 plants tested at seven times the standard dose contribute very little information. Under both the fitted logit and probit models it is extremely likely that all 100 plants would die. Because of this, we should regard prediction at, say, 4 times the standard dose as being an extrapolation.

Example: Gold grades

Figure 19 illustrates a warning about the use of non parametric procedures and transformations. It shows two log-normal distributions for gold grade. The distribution shown as a continuous line has a mean of 3. The distribution shown as a dotted line has a mean of 3.2, which is substantially higher.

A non parametric procedure might look at the medians of these distributions. The distribution shown as a continuous line has a median of 2.77 which is substantially better than the median of 2.50 for the distribution shown as a dotted line.

Similarly, if we took logarithms of the gold grades before comparing the distributions we would find that the distribution shown as a continuous line appeared to be superior.

The message of this example is that you need to remember your primary purpose as you consider alternative types of statistical analyses.

8.6 DESCRIBING THE BALANCE BETWEEN CONFIDENCE AND UNCERTAINTY

This section is concerned with the formal testing of statistical significance. This means answering questions like 'Is this model good enough?' and 'Might the size of the apparent trend be just a fluke?'

Formally, the 'statistical significance' of an effect is the probability that the observed effect would seem to be as large as it is, or even larger, under the supposition that there is really no effect at all. Intuitively, the statistical significance tells you that if an effect is a fluke then how rare a fluke is it.

The calculation of such formal levels of statistical significance is sometimes given undue emphasis. It is sometimes (incorrectly) regarded as providing the means for determining whether research results are important or significant. People sometimes consult a statistician when they are merely seeking a seal of approval for their work.

8.6.1 Statistical Significance

The testing of statistical significance means checking whether something which has been observed would be likely under some hypothesis. The hypothesis being used as a benchmark is usually referred to as a 'null' hypothesis. If the observations are extremely unlikely under the null hypothesis then it is common to say that the observations are 'statistically significant'.

In most circumstances, significance testing takes the form of comparing a summary statistic to a reference distribution.

Example: Orientations of planetary orbits

Todhunter (1949, page 222) says that the Academy of Sciences of Paris offered prizes in 1732 and 1734 for memoirs discussing the physical basis for the inclination of the planes of rotation and revolution of the six planets then known. No prize was awarded in 1732. Daniel Bernoulli shared the prize in 1734 with his father, John. At the beginning of his memoir he gave three approximate calculations suggesting that the observed alignments would be highly improbable under the hypothesis that they were randomly distributed over a sphere.

The first argument notes that the greatest inclination between the planes of revolution of planetary orbits is between those of Earth and Mercury, and is only $6° 54'$. In the terminology of this section, the maximum inclination of the planes of revolution is a summary statistic. The reference distribution of this statistic could be computed on the null hypothesis that the orbits are oriented at random. The observed value of the summary statistic is in an extreme tail of the reference distribution, so we can say that the null hypothesis can be rejected or that the data is statistically significant.

Daniel Bernoulli's computation could be improved in a few ways. It is interesting as an early example of statistical significance testing.

UPDATE BELIEFS AND UNCERTAINTIES

The yardstick of error variance

In many circumstances, statistical significance is judged by looking at the ratio of a quantity of interest to an estimate of an error variance. The details might be within a computer program, but you should at least be aware of what error variance is being used as the basis for comparison.

Consider a fertilizer trial studying the effects of various amounts of various fertilizers on the yield of a crop. If the trial is conducted in a single season then the error variance which is being used as a yardstick for judging statistical significance does not include the variation between seasons. So any conclusions about the reliability or the reproducibility of the results should admit that the season-to-season variation has not been considered.

Sometimes the populations of which the means are being compared are expected to have different variances. When the variances are regarded as known this is easy to allow for by simply calculating the variance of the difference between sample means. When the variances are regarded as unknown then a difficult-to-use test called the Behrens–Fisher test is recommended.

Example: Comparing gold assaying procedures

The grade of a mineral sample was determined in a routine way. The result was 2.3 g/tonne. Several months later, several other determinations of grade were made using a different assaying procedure. These gave the results 3.2, 3.01, 3.82, 2.59, 3.14 and 2.87 g/tonne. Is the difference between the average grades obtained using the two procedures statistically significant?

An inappropriate procedure that is sometimes used in this situation is a one-sample t-test, checking whether the mean assay with the second procedure is significantly different from 2.3. The t-statistic for this test is 4.78 with 5 degrees of freedom which gives a two-tailed significance level of 0.01. This indicates that the difference is statistically significant, but this conclusion is irrelevant because it is not answering the correct question.

The number 2.3 is not the true mean assay result for the first procedure. It is merely a single measurement. This makes it the sample mean for the first procedure, but it is not the population mean.

An appropriate procedure for testing whether the average of the six assays with the second procedure is significantly different from the single assay with the first procedure is a two-sample t-test, in which one of the samples happens to be of size 1. The two-sample t-statistic calculated in the usual way is

$$t = (\bar{X}_1 - \bar{X}_2)\sqrt{\frac{n_1 n_2 (n_1 + n_2 - 2)}{(n_1 + n_2)[(n_1 - 1)s_1^2 + (n_2 - 1)s_2^2]}} = 1.807.$$

Care must be taken in evaluating this expression because the sample variance of the first sample, s_1^2, is undefined. The product $(n_1 - 1)s_1^2$ should be taken to be zero. The t-statistic has five degrees of freedom, so the two-tailed significance level is 0.13.

8.6.2 The Correlation Coefficient is not a Good Yardstick

Some people have somehow gained the impression that the correlation coefficient and the multiple correlation coefficient, R^2, are useful as yardsticks in the sense that a model with a higher R^2 is better than a model with a lower R^2.

This is not necessarily true. Looking at error variance is a much better way of comparing models.

Example: Fat depth

A researcher was investigating the relationship between measurements of fat depth made on live sheep and the amount of lean meat on those sheep. The amount of lean meat was determined by slaughtering the sheep in the usual way then cutting the meat from the carcasses particularly carefully.

The researcher was comparing two regression equations.

- One predicted the percentage of lean meat using carcass weight and fat depth as predictors. It included a quadratic term in fat depth. This equation had a multiple correlation coefficient of $R^2 = 0.5$
- The second regression equation predicted the mass of lean meat, again using carcass weight and fat depth as predictors. It included terms such as the square of carcass weight and the product of carcass weight times fat depth, which made it harder to interpret, but it had a multiple correlation coefficient of $R^2 = 0.95$

The researcher wanted advice. Should he chose the intuitively sensible regression equation or the 'better' predictor?

The researcher's only problem was that of thinking that R^2 is worthy of attention in this context. Working with the percentage of lean meat rather than with the mass of lean meat as the response variable removes much of the effect of carcass weight. It is not surprising that the first regression equation explains a smaller proportion of the variability – there is less variability to be explained.

A better way to compare the two regression equations is to compare the mean squared errors expected when they are both used to predict the same quantity. For instance, the second equation can be used to predict the percentage of lean meat by dividing by the carcass weight. When the regression equations were compared in this way, neither was clearly better than the other. There were some regions where each appeared to be slightly superior. The researcher decided to use the 'intuitively sensible regression equation', which was the first one.

Comparing methods of measurement

People are often interested to compare methods of measuring what is essentially the same quantity. This commonly takes the form of comparing an expensive, accurate, slow measurement with a cheap, imprecise, quick measurement. Let's refer to them as X_1 and X_2, respectively.

The correlation between X_1 and X_2 depends crucially on the range of values. If the range is large then correlations such as 0.99 are common, but should not be the basis for confidence that the cheaper measurement can be satisfactorily substituted for the more expensive measurement. If the range is small then correlations will be smaller, which sometimes leads people to suspect that their procedures are not as good as those of other researchers who obtained larger correlations. Such use of the correlation coefficient is inappropriate.

One good simple way to look at the tightness of the relationship between X_1 and X_2 is to look at the mean and variance of $(X_1 - X_2)$. A graphical display of the difference $(X_1 - X_2)$ against the average $\frac{1}{2}(X_1 + X_2)$ is very useful for showing outliers or regions where the relationship between X_1 and X_2 might be treated differently. See Bland and Altman (1986) for some examples.

8.6.3 Does the Model Fit?

There are many different models which could be fitted to response surface data. Even for a single factor the possible models include linear, quadratic, cubic, quartic, multiplicative, exponential and logistic. With more than one factor we may need to make choices about excluding some factors and including some interactions as well as choosing univariate functional forms.

Sometimes background knowledge suggests that a particular functional form is likely to be appropriate, but often a pragmatic approach requires simply that we select a model which is a good fit. We would like to know whether there might be some other model which provides a substantially better fit to some data. For a particular postulated model, we need to ask 'Does the model fit?'

In order to answer this question we need to get some information about the amount of noise in the data. Unless we already have such information, it is necessary to repeat at least some of the runs in the experiment.

The commonest ways of arriving at an expected value for the residual mean square are as follows.

- If a series of similar experiments are conducted then the residual mean squares can be monitored to gain an idea about the general level of residual mean squares. (The amount of noise associated with data of the particular type).
- Replicates of experimental runs give information about the amount of noise associated with experimental procedures.

Example: Simulated data (continued)

Continuing the example using the simulated data shown in Table 1 on page 211: An estimate of the standard deviation of the residuals is 8.82 units of yield.

The term in $(A - 50)^2$ has a coefficient which is small compared to its standard error, so we cannot have much confidence about the location of a maximum and should endeavour to get more information about the way in which the yield varies with A.

If the response surface was a good fit then we would expect that the residual mean square (77.83) would be consistent with the hypothesis that the squared differences between replicates would average twice the residual mean square, because the variance of a difference of independent quantities is the sum of the variances. In fact, the average squared difference between replicates is 6.375 and this is nowhere near 2×77.83. Hence we conclude that the model is not a good fit.

8.7 USING COMPUTERS FOR STATISTICAL ANALYSIS

This section discusses some issues about how computers can be used for statistical analysis. It is assumed that readers will be using computers to help analyse experimental data. This is a complicated topic. It will suffice here to make a few points.

Proceed in an auditable way. My preferred method of analysing data is to run an edit session in one window and to run a statistical package in another window. I create a list of the commands that I find useful in the edit window while trying them out in the window running the statistical package. In this way, I maintain an audit trail of my analysis.

In particular, I try not to modify the file containing the original data. If a data point seems to be incorrect then I overwrite its value using a command within the statistical package.

I was most conscious of doing this when potentially an expert witness in a court case.

Choosing a computer package (or packages). It may be appropriate to use different computer packages for different stages of data analysis. For instance, I consider some packages to be better for initial data manipulation and preliminary analysis while others are better for definitive analysis. Presentation quality graphics might be produced with a package which was not used for formal data analysis. The disadvantages of using more than one computer package are that you need to become familiar with how to use more computer software and that there will generally be some effort to be expended converting data between different formats.

What does it do? If you want to know what a piece of computer software does, whether it gets the right answers and how to interpret its output, then there is one very good way to start. You should take a data set from your favourite textbook which is as similar as possible to the data which you wish to analyse, have the computer analyse this data and compare the answers given by the computer with the answers given by the textbook.

Don't get lost! Remember that computers only do what you tell them to do. Many of the steps which you would do if you analysed an experiment by hand tend to be neglected if data is analysed using a computer, because the computer has not been told to do these things.

Pay attention to the parts of the computer output which monitor what has happened – how many records have been read, what files were used for reading and writing?

Ask what parts of the computer output matter to you. Remember that computer software is written to satisfy a wide variety of people analysing a wide variety of experiments with a wide variety of objectives. Many statistical tests are included because they are likely to be useful to other people, experiments or objectives.

Use graphics. Sometimes you will use data analysis techniques which are not appropriate, do not tell the whole story, or need to be carefully monitored. Humans are generally much better at interpreting visually than at interpreting numbers.

8.8 WHEN TO SEEK HELP WITH STATISTICAL ANALYSIS

It is easy to know to seek help with statistical analysis if complicated procedures seem unavoidable. This section discusses a few of the most common reasons why experimenters ought to seek help with the analysis of a particular experiment, even though they may have competently analysed much experimental data in the past.

8.8.1 Beware of Experiments Conducted as Split Plots

When experiments are conducted as split plots, with hard-to-adjust factors being varied relatively infrequently and easy-to-adjust factors being varied relatively often, the estimated error variance which should be used as the yardstick for checking the statistical significance of the hard-to-adjust factors is not the same as the estimated error variance which should be used as the yardstick for checking the significance of the easy-to-adjust factors.

8.8.2 Beware of High Correlations between Estimated Coefficients

Many statistical computing procedures report a matrix of correlations between the estimated coefficients of a fitted model. If correlations are near to either $+1$ or -1, then compensating changes could be made to the estimated coefficients with little change to the fitted model at the places where you have data. This may mean that your model is more complicated than can be supported by the available data.

Sometimes the correlations can be reduced by adding or subtracting constants from the explanatory variables (for instance, by using $T - 220$ as an explanatory variable rather than using T, when T is the temperature in °C for a factor with levels 205°C, 220°C and 235°C.) Such adjustment to variables is generally helpful.

8.8.3 Beware of Looking at the Most Statistically Significant of a Large Number of Models or a Large Number of Comparisons

Beware of analyses in which you have implicitly considered a very large number of possible regression equations (say, more than 30) and chosen the best. The selection biases the estimated regression coefficients, overstates the statistical significance of the equation chosen and underestimates the error variance. The solution is to think before collecting data and to only do those analyses which you had considered likely to be interesting before you saw the data.

You should also be wary of formulating a hypothesis and testing it on the same data set. This is more often a problem with analysing observational data than with analysing experimental data. For further information about this issue, see Miller (1990).

8.8.4 Beware of Fitting Models if you do not Understand the Criterion for Fitting

Various criteria for fitting models are discussed in books which are more theoretical than this one. A common criterion for fitting models is to minimize the sum of

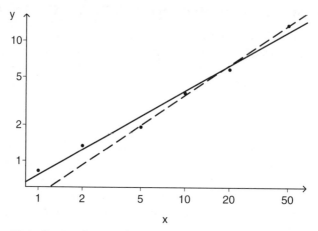

Figure 20. Plot of some data on logarithmic scales. The continuous line minimizes sums of squared residuals on the logarithmic scale. The dashed line minimizes sums of squared residuals on the raw scale

squared deviations between the data and the model. These deviations are referred to as residuals, and the method is referred to as 'least squares'.

The only notable weakness of least squares is that it is not robust in that an estimate of a parameter can be greatly influenced by even one outlying value. However, the non robustness is easy to understand and to make allowances for.

Many common statistical procedures can also be justified on the criterion of 'maximum likelihood'. This means that those procedures find the estimates of parameters which make the data as likely as possible. Generally, it is essential to assume that errors are normally distributed in order to show that such procedures are optimum according to the criterion of maximum likelihood, whereas the criterion of least squares does not require such an assumption. I tend to think in terms of least squares so that the procedures seem more robust.

Even the criterion of least squares needs to be restrained by intuitive understanding. Consider the following example.

Fitting on a logarithmic scale. Consider fitting a model to the following data.

x	1	2	5	10	20	50
y	0.83	1.34	1.93	3.68	5.84	13.56

Figure 20 shows the data and two lines of best fit, all on a logarithmic scale. The best fit on the logarithmic scale is

$$\log(y) = 0.70 \log(x) - 0.27.$$

This is shown as a continuous line on Figure 20. This minimizes the sum of squared residuals on the logarithmic scale, but doesn't minimize the sum of squared residuals on the scale of the raw data. That is minimized by the equation

$$\log(y) = 0.83 \log(x) - 0.65$$

# UPDATE BELIEFS AND UNCERTAINTIES

which is shown as a dashed line. As can be seen in Figure 20, this line has a smaller fractional error for the points which correspond to large numbers than to the points which correspond to small numbers.

Generally, the continuous line is likely to be preferred because residuals of equal size on the logarithmic scale are equally important.

9

Revisit the Objective

After each trial reconsider your objective in the light of the results obtained. Does it still seem to be sensible? Also reconsider your experimental strategy. Have goals been achieved? What more needs to be done? Is further experimentation justified?

At the end of your experimental programme, you need to report your results and communicate the conclusions which can be drawn. However, your task should not be regarded as finished with the completion of a report. In order to maximize the impact of your work, you should make the effort to take whatever additional steps are necessary to help make changes happen.

9.1 RECONSIDER OBJECTIVE AND PLANNED STRATEGY AFTER EACH EXPERIMENT

This section is relevant to the situation where one experiment has been completed and further experiments had been planned.

9.1.1 Does the Objective Need Reconsidering?

Sometimes the results from an experiment cast a new light on the objective of the overall experimental programme.

- Does it still seem sensible?
- Is it still worthwhile?
- Should other response variables be considered?

9.1.2 Does the Strategy Need Reconsidering?

The result of an experiment may indicate that a change in experimental strategy would be more fruitful.

- Do the assumptions on which the strategy was based still seem to be reasonable?
- Are some conclusions in conflict with the conclusions of earlier experiments?

Perhaps it is time to stop, either to abandon the research thrust because further investment of resources does not seem justified; or to conclude the research thrust with some degree of satisfaction because further refinement of the conclusions would not be justified.

Monitor the estimates of the components of noise. If they have changed substantially then some reaction may be required. Differences in experimental technique, differences due to measuring equipment, and many other factors might be found to affect variability – and this knowledge can lead to quality improvements:

- change numbers of runs in future experiments;
- change experimental techniques.

It may be desirable to swap to another of the types of experimentation discussed in Chapter 4, namely studying sampling and testing errors, main effects experiments and response surface methodology. The smoothest progression is to start by studying sampling and testing errors, then proceed using main effects experiments to find out what factors are important and later use response surface methodology to gain an adequately precise knowledge of the effects of the most important factors. However, it is very common for an experiment conducted assuming that sampling and testing are already understood to lead to the conclusion that sampling and testing are not adequately understood. And it is also common for results and observations on experiments seeking detailed knowledge about some factors to suggest that other factors should be considered.

Example: Apple damage from transport in wooden bins

Engineers from the New Zealand Department of Scientific and Industrial Research had, jointly with the New Zealand Apple and Pear Board, conducted several trials that investigated damage to apples caused by transport in wooden bins. These trials had made extensive engineering measurements, but other aspects of the tests had not been systematically planned. The trials involved several varieties of apples at different maturities and damage was measured in a variety of ways.

Following its experience with these earlier tests, the New Zealand Apple and Pear Board insisted on the involvement of a statistician; this led to the trial that is described in Maindonald and Finch (1986). John Maindonald has provided me with some additional material about the circumstances of this trial.

The factor of primary interest was the difference between mechanical suspension and air suspension on trucks. A factor of secondary interest was the position of the wooden bin on the truck. Twelve positions were compared. Another factor of secondary interest was the direction of travel. The trucks travelled between Henderson and Hastings. The direction of travel was expected to affect the gradings of apples because apples which travelled to Henderson were graded at Henderson using graders based at Henderson whereas apples which travelled to Hastings were graded at Hastings using graders based at Hastings. Several other factors were held as constant as possible.

- The two trucks used in the trial were the same age and model.
- They travelled together, taking turns at leading.
- All apples were Granny Smiths, taken from the same region of a coolstore.
- The same graders were used at Hastings on both occasions. However, the graders used at Henderson were not the same.

The trial did not provide evidence that air suspension was better than mechanical suspension. Variation in bin stability made it impossible to detect the much smaller differences that might have occurred due to differences in the suspension system. However, the trial did provide evidence that rail transport was superior to the use of road transport when bins were not properly stabilised. Note that the researchers might have been alerted to potential problems with unstable bins had they talked to orchardists about transport damage.

9.2 COMMUNICATE RESULTS

There are several jobs to be done which might collectively be referred to as 'reporting results' or 'communicating conclusions'. A brief list follows.

1. Summarize the results of individual experiments in a technical way.
2. Sell the conclusions drawn to the people who ought to be convinced to change some aspect of their activities.
3. Provide feedback (or even complain) to the people responsible for setting objectives. They may not have fully comprehended the consequences of their decisions.
4. Write down suggestions for future experimentation.
5. Write down suggestions for the monitoring of processes which have been investigated and information relevant to the setting of specifications.

9.2.1 Focus on the Interests of your Audience

The key steps are to think about who your audience is and to think about the purpose of the communication.

Who is the audience?

Presentation for managers, peers, subordinates and outsiders each require special consideration.

- Managers: be brief. Give a thorough introduction and make major conclusions very explicit. Be sensitive to politics. Be convincing.
- Peers: give a fair and detailed account of both practical and theoretical aspects. Consider your peers in the future. It may be desirable to record your bad results somewhere else.
- Operators: emphasize practical implications and indicate ways in which practices might still need modification by people with on-the-job experience. Why should they believe you?
- Outsiders: indicate implications relevant to intended audience. Also, be careful about ways in which the organization might be sensitive to public opinion.

What should be reported?

A few specific points are as follows.

- Explain the background and objectives of the experimental work.
- A complete description of the experimental runs and the results recorded should be stored. Details should be sufficient to enable the experiment to be repeated or to be

re-analysed. The place of storage should be a file or a library, not an out-of-the-way cupboard. The period of storage should comfortably exceed the period over which the data might prove useful.
- Use graphical methods whenever this will aid comprehension. If there is some other simple way of presenting your conclusions, use it.
- State any assumptions, limitations to the scope of the conclusions or other uncertainties simply and honestly – this will help your credibility.
- When presenting statistical analyses, remember that statistical significance is not the most important feature. It is far more important that the results and the substantive significance be readily comprehended.
- Make clear which information is directly relevant to the most important problem, which information is only being presented for the sake of completeness, and other levels of importance.

9.2.2 Expressing Conclusions in Words

There are many books about technical writing, presentation of training courses and similar topics. Different people find different books to be helpful. I suggest that you consult some colleagues and a librarian for a short list of suggestions if you would like to choose such a book. My versions of some of the points made by such books are listed below.

I recommend Halmos (1970) on the more specific topic of how to write mathematics.

Communicate in a way that your audience will understand. This means emphasising whatever your audience is likely to be most interested in: the purpose of your experiments, your conclusions, what needs to change, the benefits, the barriers, or the technical details relevant to replication of the experiments. It may mean restricting the amount that you say or write. It may mean using diagrams, photographs or demonstrations, as well as words. It may mean avoiding jargon, or it may mean deliberately using the same jargon as your audience.

Have a clear, logical structure to what you say or write, and explain that structure to your audience. Particularly if you are describing a sequence of experiments rather than a single experiment, the structure is likely to be richer than to simply use the headings 'Aim', 'Method', 'Results' and 'Conclusions' that I was taught to use in secondary school. Your audience should know at all times whether a paragraph is describing intentions, plans, actions, observations or inferences.

Communicate your value judgements. Tell your audience which parts of your presentation are objective and which are subjective, personal or idiosyncratic. Tell them what you regard as important and what you regard as unimportant. And openly state any uncertainties, qualifications or restrictions to the area of validity of your conclusions.

Some people seem to believe that science should be value-free and therefore try to avoid communicating their value judgements. I disagree. In my experience, those same people are generally willing to distinguish good and bad theories, good and bad methodologies, and promising and not-so-promising lines of enquiry.

Make linkages clear. The hardest part of the task of interpreting words is working out the linkages which the writer of those words was trying to convey. When writing, you should try to make this task easier for your readers. Some hints are as follows.

- When you wish to refer to something which is the same thing as you referred to previously, first consider using exactly the same words. Synonyms, abbreviations or pronouns should only be used when working out their meaning will not distract readers.
- Don't use the same set of words to mean two different things.
- Use the word 'the' carefully. It should not be used in an attempt to make something seem more important. Whenever you write 'the X' it implies that readers should be able to work out which specific X. This often means that there is some linkage between the 'X' here and an 'X' somewhere else.
- Use displayed lists or flowcharts, in preference to long paragraphs of text when many items are linked as a group, as a sequence, or in a slightly more complicated way.
- Be careful when using qualifiers like 'all', 'any', 'every' or other words describing the scope of a statement. If the meaning of a sentence is not completely unambiguous, reconstruct it.

Have a policy about redundancy. People often understand and remember better if explanations can be repeated, preferably in a variety of ways. Yet conciseness is also a virtue, so you have to make a judgement.

Read your own writing. I find it useful to try to read aloud what I have written, preferably after letting the relevant ideas settle for at least a couple of days. This checks that the logic behind the words can be extracted easily. Note that this process is different from proofreading to check for mistakes.

Many people have been trained to write about experiments in the passive voice and to avoid the use of first person pronouns. I think that this training is a barrier to clear communication, but I am not foolish enough to believe that it will be stopped because of my opinion. Rather, I hope that some people will read their own writing aloud to themselves and thereby avoid some of the most clumsy and pretentious constructions which result from efforts to write in the passive voice and avoid first person pronouns.

Get help with grammar, style and spelling, if you need such help and the effort of detailed redrafting is justified. For most communication about experiments, I believe that detailed redrafting is not justified. The potential benefit is small because a well-structured document written with poor grammar, style and spelling can be understood without excessive effort. And the effort required is great, because many adults feel emotionally uncomfortable about accepting help to improve their writing style.

9.2.3 Presenting Tables

Sometimes, detailed tables of information are included in appendices for the sake of completeness. A better option is often to describe the location of a computer file where the information can be obtained. In this section I am not concerned about such tables.

Rather, I am concerned with tables which are presented as part of an attempt to sell a message following a programme of experimentation.

Ehrenberg (1977) gives some excellent advice on the presentation of tables, and gives comments on some of the most common objections to his recommendations. My version of his ideas as they apply to reporting of experimental data, analysis and conclusions is as follows.

- Tell your readers what patterns and exceptions they should look for in a table. Then these patterns and exceptions should be obvious at a glance.
- Use as few significant digits as possible, in order to simplify the memory and mental arithmetic tasks required for comparing numbers. (The complete results of your experiments, without rounding, and your very thorough analysis of that data should be recorded somewhere, but you don't have to bore everyone with the minute details.)
- Consider including averages, totals, percentages or other figures to help readers distinguish typical trends and exceptions.
- In deciding on overall layout, remember that it is easiest to compare numbers which are in the same column and near to one another.
- Choose the sequence of rows and columns carefully. For instance, if there is a column for each of a large number of chemical constituents then order the columns in a way which will help readers to see the patterns and exceptions. This might mean grouping heavy metal oxides together, grouping impurities which arose from the same source together, or putting constituents in decreasing order of abundance. Alphabetical order is often not the best option.

9.2.4 Presenting Graphs

Graphs should be used to visually support your conclusions and explanations. Cleveland (1985), Tufte (1983) and Wainer (1984) are useful sources of ideas about the principles of presenting graphs. Some personal comments are as follows.

- Show the data. Don't use chartjunk. Try to make people remember the message behind the data, not to win a prize for artistry.
- If trends and exceptions are clear on a simple graph, then that graph will be effective. It is possible to put a large amount of quantitative information onto a graph. If readers know that important messages can be ascertained by studying a single graph then they will be willing to study that graph carefully.
- Make the task of interpreting graphs as simple as possible. This tends to be compromised when computer software is used for drawing graphs, because it is very difficult to write software to perform some tasks. When several graphs are to be compared, use the same scales (or as similar as possible). Label legibly, correctly and unambiguously. Use upright text whenever possible. Put labels on lines in preference to using a legend, because it makes the task of interpretation easier. If you must use a legend, try to simplify the memory task of people looking at trends and exceptions by choosing line or point types which have some natural relationship to the distinction between them.
- Be honest with the visual metaphor. Don't use multiply labelled axes or complicated graphs. People usually trust graphs, but that trust is fragile.

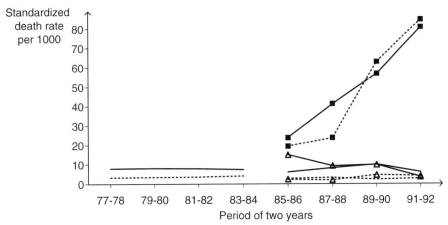

Figure 1. Annual death rates per 1000 haemophilia patients. Data points for severe haemophilia are linked by solid lines. Those for moderate and mild haemophilia are linked by dotted lines. Data for patients known to be HIV positive are shown as solid squares. Data for patients known not to be HIV positive are shown as open triangles. Data for patients whose HIV status is unknown are not plotted as points. Only lines joining these points are shown.

Before 1985 it can be assumed that virtually all patients were HIV negative

If there is a substantial risk that readers might wrongly interpret a scale as starting from zero then draw their attention to the origin or a break in the scale. When using a logarithmic or other transformed scale, label it in as meaningful a way as possible. Make legends comprehensive and informative.
- Pie charts have the good feature that they let readers know that the quantities being displayed are proportions. They have the disadvantage that people are not good at judging the relative sizes of wedges with different orientations.
- Colour is useful in graphs only for attracting attention and for distinguishing between a small number of categories.
  Note that the colour of coloured lines becomes much less obvious as a graph is reduced in size. Do not use colour on graphs if they might need to be photocopied.
- Proofread graphs just as you would text. Check that visual clarity is maintained under reduction and reproduction.

Example: HIV and haemophilia

Figure 1 is similar to Figures 1(a) and 1(b) from Darby et al. (1995) and is based on the data given in that paper. A virtue of this graph is that it shows an important message in an undisguised way. The change in death rate for haemophiliacs from before 1985 to after 1985 appears to be mainly due to HIV. The death rate for HIV negative patients has not changed, but the death rate for HIV positive patients has dramatically increased with the death rate being little affected by the type of haemophilia.

9.3 HELP MAKE CHANGES HAPPEN

Why was the experiment conducted at all? Take some personal responsibility for helping the original purpose to be achieved. You may be in a good position to help explain what might be achieved by change.

Provide any feedback which might be useful.

- Was the problem adequately specified?
- Are there aspects of the problem that other people have tended to ignore?
- What problems might be worked on next?

9.3.1 Why did you do the Experiment in the First Place?

What was your original purpose? What can be done? What might be done better? What needs to change?

9.3.2 What are you going to do Next?

Take personal responsibility for ensuring that improvements are implemented. You may also be able to advise about how to monitor the implementation.

Example: Trial of steering strategies

Many years ago (and this story might contain more mythical elements than many of my other stories) an experiment was done to investigate the effect of small amounts of alcohol on people's performance at steering a vehicle through a course delimited by witches' hats.

The actual results were not regarded as surprising, except that a very senior policeman who was on an advisory committee thought that the average performance was surprisingly poor, even for the runs when the human subjects had not ingested any alcohol. The senior policeman asked who the subjects were, and was informed that they were the usual mix of mostly university students and housewives.

He offered to provide police officers as subjects for a repeat of the experiment, thinking that for such competent, experienced drivers their performance on the steering task might be less affected by small amounts of alcohol.

What happened in the repeat of the experiment? The police did not perform as well as the university students and housewives. Anecdotal evidence suggests that at that time police were taught never to allow their hands to cross over while steering a vehicle and that this training had substantially inhibited their performance on the steering task used in the experiment.

What happened to the results from the experiment? They were suppressed. Evidence that police officers were poorer drivers than university students was not politically acceptable.

What would you do if you were the experimenter in this situation?

- Performance on the artificial tasks of the experiment is not a particularly good measure of driving performance on the road, so perhaps the suppression is acceptable?
- Perhaps there is evidence here that police driver training could be improved?

REVISIT THE OBJECTIVE

- Maybe you have strong views about openness and honesty in public services or other organizations?
- Or perhaps you think that this is none of your business?

It is important to make a careful assessment of what might be achieved, both changes for the better and changes which make matters worse, by making some people over-defensive or by making your own situation uncomfortable. There might be strategies available to enhance the changes for the better or to minimize the undesirable changes.

9.4 CONTRIBUTE TO IMPROVING OTHER THINGS

Having finished a programme of experimentation, one responsibility for action has been discussed in the previous section. It is to help make changes related to the primary purpose of the experiments. Other changes may be suggested by the outcomes from your experiments. You should be prepared to make some effort about them also.

9.4.1 Changes to the Experimentation Checklist

Some users of my experimentation check-list might find that they wish to tailor the check-list to their specific needs. Consider whether the steps could be changed, re-ordered or improved for the types of experiments that you do.

9.4.2 Improve the Statistical Thinking of other People

By 'statistical thinking' I mean the expenditure of thinking effort on worrying about variation and uncertainty.

- Sometimes the subject of primary interest is variation. For instance, genetics describes variation and make predictions about variation. Geneticists were the first scientific group to be interested in statistical methods because they had a natural interest in variation. In contrast, many engineers and physicists seem to think that any consideration of variability is an admission of weakness.
- Sometimes it is possible to predict how much variation there will be. For instance, radiation counts from the decay of radioactive isotopes or from X-ray fluorescence have variations which is accurately predicted by the Poisson distribution. We are also able to make predictions about the accuracy of carbon dating techniques.
- Sometimes the amount of variation can be approximately predicted from experience with similar processes. For instance, the amounts of sampling and testing variation can often be predicted. Knowledge of the sizes of the components of variation which are likely to arise from various sources can be useful for choosing between different sampling and measurement strategies and for predicting the precision which is likely to be achieved by a given amount of experimental effort.
- Discussion and assessment of uncertainties is important in deciding on experimental strategy.

You may be inclined to encourage other people to take a less deterministic view of the world.

9.4.3 Provide Information for Future Experimenters

You should record the complete plans and results of your experiments in some way so that they can be accessed in the future. A summary of outcomes, the precision achieved and an indication of any problems with protocols or experimental methods should be made conveniently accessible to your current and future colleagues.

9.4.4 Lessons for Managers of Experimenters

Managing people whose job includes generating new ideas and questioning old ideas seems to be very difficult, or at least to require an appropriate management style. One manager commented to me 'It's like herding cats!' I cannot claim expertise at managing people. I merely have some comments.

Managers of experimenters may include people who gave a broad specification of the original problem. They need to consider feedback about how the experiments proceeded, so that they can contribute to discussions about which problems might be undertaken next.

Experimentation is a risky business. People conducting experiments must be expected to spend much of their time doing work which in hindsight might later be judged to have been unproductive. They should not be held individually responsible for short-term results. However they can be held responsible even in the short term for using sensible processes to conduct experiments.

In judging plans for proposed experiments it is important to have a bias towards 'giving things a try'. The costs of experiments are much easier to quantify than the benefits. However, this should not excuse experimenters from being conscious of economics.

9.4.5 Lessons for Trainers of Statisticians

I believe that the training of statisticians is not well matched to the contributions that many statisticians make after they settle into the non-academic world. This is not an idiosyncratic view, but is widely shared. For instance, Hunter (1981, pages 113–4) wrote

> The statisticians' training, narrow and technical, is the orderly climb up a staircase of mathematical problems that each have only one right answer. Later steps rest on earlier ones. Progress is always up. Teachers watch the climbing steps of the fledgling statisticians, and help them to master the steps, one at a time. The path is clear. The atmosphere is safe (one, after all, is indoors).
>
> Statisticians' work, for which this training is supposed to equip them, is the disorderly climbing of rugged hills, outdoors, in fair weather and foul. The path is anything but clear. A promising path can get lost in the tangled undergrowth or a patch of dense forest, or it can plunge over the edge of a ravine. (Ravines tend to appear at the most inconvenient times.) ... The statistician, as consultant, is often asked to act as a guide for such expeditions into unknown territory. Some graduates find this activity exhilarating. Some find it otherwise. All find that their world has changed.

Hunter may have been inspired to use such entertaining language by some of the colourful expressions used by Salsburg (1973), on which he was commenting. Salsburg's view is that 'We live not in a time of information explosion but in a time of data inundation.' Statisticians can contribute by 'attempting to stem the tide of confusion', 'asking stupid questions', coming to conferences 'with a non-deterministic outlook' and 'trying to make sense out of data'. Many statisticians do this, despite their training.

In my view, a key to changing the training of statisticians is to change their assessment. The technical, mathematical part of the training of statisticians is already sensibly assessed using examination questions which have right answers. Other parts of the training of statisticians could be assessed using examination questions which require descriptive answers, like questions on history or philosophy.

For instance, an examination question might take the following form.

> Two possible experiments have been proposed for investigating ... . Some background information is as follows. ... Some details about these proposed experiments are as follows. ... Discuss this proposed experiment, including the following issues.
>
> 1. What teamwork, personnel and management issues might be relevant?
> 2. Discuss the goal of the experiment and how it might be made more specific.
> 3. Speculate on experiments which might have preceded or which might follow the experiment discussed.
> 4. Consider the adequacy of the measurements made. Might different measurements be used at other stages of the experimental programme?
> 5. What are the good features and potential weaknesses of the proposed experiments?
> 6. Should any aspects of the experimental procedures be spelled out in greater detail?
> 7. What alternative experimental designs might be considered?
> 8. If a meeting were to be held to discuss some options for further experimentation, collection of routine data or major capital investment, what issues would need to be discussed at that meeting?

Glossary

Accuracy An 'accurate' measurement procedure is one which gives results which are generally close to the true value. This is sometimes contrasted with a 'precise' measurement procedure which gives consistent results which may not be clustered around the true value. Because the term 'accuracy' includes both smallness of bias and good precision, it is often clearer to make separate, explicit statements about bias and precision.

Analysis of variance This describes the total variation in a set of data as having components which are attributable to various sources of variation. Some of these sources may be controllable factors, in which case the corresponding component of variation can be used to check whether the apparent effects of those factors are statistically significant. Other sources of variation may be interpretable as summarizing components of experimental error, the lack of fit of some model, or inconsistency in the effect of a factor over the range of the conditions of the experiment.

Average The average (or mean) of a set of numbers is simply the total divided by the number of items contributing to that total.

A common source of confusion with the use of the term 'average' is that the set of things being averaged is not completely unambiguous. In particular, it may be necessary to distinguish between a sample average and a lot average. A common notation is to denote an unobservable, true average (such as a lot average) by the Greek letter μ (mu), and to denote an observable average by a Latin letter with a bar (e.g. $\bar{X}$).

Bias A bias is a discrepancy between the average that you would get from a large number of experimental determinations and the true average that you are interested in estimating. Note that biases are seldom constant, and attempts to estimate the average bias and correct for it seldom turn out as well as people hoped.

Block A block is a set of runs or trials which are likely to be more homogeneous than the full set of runs or trials, not because of the experimental treatments applied being similar but because other sources of variation are likely to be similar.

Blocking Blocking is said to be used when an experiment is organized into blocks of experimental runs or trials. Within-block comparisons of treatments are generally more informative than between-block comparisons because of the smaller amount of variation within blocks.

In a test of tyres of various makes, sets of four tyres of different makes were put onto cars. Each three months, the tyres were rotated. After a year's driving, the wear on each tyre was measured. The four tyres on a single car are more directly comparable than tyres on different cars because they have been used in a more similar fashion.

Confounding One effect is said to be confounded with another if the two cannot be separated. This may not matter if neither of these effects is of substantive interest.

Continuous variables These are quantities which can take values on a continuous scale. When the factors in an experiment are continuous variables, such as concentrations of chemicals, dimensions, temperatures and times, then the levels used in an experiment can be chosen freely, and will usually be chosen so that the range used in the experiment is of particular interest.

Contrasts A contrast is a linear combination of the parameters of a model such that the sum of the coefficients in that linear combination is zero, or an estimate of such a quantity. Contrasts used in practice are generally combinations of differences between treatments which have some special interest.

For instance, if a chemical trial is conducted at 110°C, 120°C, 130°C and 140°C, giving average responses $y(110)$, $y(120)$, $y(130)$ and $y(140)$, then the linear combination

$$0.03y(140) + 0.01y(130) - 0.01y(120) - 0.03y(110)$$

is an estimate of the average change in response per degree Celsius over the range of temperatures tested.

Correlation A positive correlation between two quantities is a tendency of their values to both be large or to both be small. A negative correlation is a tendency for one of their values to be large when the other is small.

The partial correlation between two quantities given a set of other variables describes such trends after making allowance for the values of the other variables.

Covariates Covariates are additional information recorded for all runs which might be used to help to explain some of the variation between experimental results, thereby reducing the residual error variance and increasing the sensitivity of the experiment.

A measurement of the quality of raw materials might help to explain differences between runs of some process. Measurement of the ambient temperature or humidity might also be useful. In an agricultural trial, the yield of some crop sown the season before an experimental crop might be a good indicator of the relative fertility of the experimental plots, and so help explain the differences between experimental results.

GLOSSARY

Degrees of freedom The number of degrees of freedom is the number of independent ways in which something can vary with the variation being assigned to a specified cause. It is often easiest to think about it as the difference between the total number of ways in which something can vary and the number of ways in which the thing can vary but the variation would be assigned to some other cause.

For instance, the sample variance of data $X_1, X_2, \ldots, X_n$ is said to have $n-1$ degrees of freedom because there are n total ways in which the data can vary, but one of these ways (namely all of the data being increased or decreased by the same amount) is assigned to the overall mean of the data.

Another example is a three-level qualitative factor is said to have two degrees of freedom because there are three ways in which the averages of the categories can vary, but one of these is assigned to the overall mean.

Extending this example we have that the interaction between two three-level qualitative factors is said to have four degrees of freedom. There are nine ways in which the averages of $3 \times 3 = 9$ categories can vary, but one of these is assigned to the overall mean, and two are assigned to the main effects of each of the two factors.

Discrete variables These are quantities which can only take defined values. The values may be unordered, such as the classification of cattle as calves, heifers, cows, bulls and steers, or ordered, such as the rating of pieces of wood for structural integrity on a scale from zero to eight according to well-established guidelines. There may be as few as two values, such as 'yes' and 'no' or 'dead' and 'alive', or as many as a countably infinite number of possible values, such as the number of insects of a particular species in a defined region.

Evolutionary operation (EVOP) This is the conduct of experiments using full-scale facilities while expecting that the process output will remain useable or saleable. Compared to experiments which are conducted under other circumstances (off-line), the ranges used for factors are usually smaller but the numbers of replications are often larger.

Experimental error The term experimental error is often used in statistical literature to describe the expected random variation in experimental results. However, the term is also used to mean a gross mistake, a particular experimental outcome for which the departure from the trend indicated by other outcomes or indicated by some theory is much greater than would normally be expected.

Experimental runs These are the units from which experiments are built. An experimental run is specified by giving the 'treatment combination' or specification of levels for all of the factors being investigated, specifying the things to which the treatments are applied, and specifying the environmental conditions and measurement systems.

Factor A factor is a variable which is deliberately varied in an experiment so that its influence may be investigated. Generally factors are variables which you have control over, and in this case they are often referred to as 'independent variables'. Factors may be continuous or 'quantitative', such as dimensions, times, temperatures and concentrations, or they may be discrete or 'qualitative', such as choices of operators, machines, tools or raw material sources.

If something seems to be a factor, but you don't have good control over it, then regard its intended value as a factor. For instance, you might wish to control temperature or flow rate but be having difficulty. You would like to regard them as factors because they influence the outcome of the process that you are investigating. The solution is to regard intended temperature and intended flow rate as factors. Actual temperature and actual flow rate might be usefully measured as well as the response of primary interest.

Foldover designs Given a two-level fractional factorial design, the foldover of that design is found by swapping all plus signs for minus signs, and vice versa.

Factorial designs These are where all combinations of levels of a number of factors are used. Sometimes factorial designs are augmented by some extra points in order to fit a response surface.

Fractional factorial designs A fractional factorial design is one in which only some of the runs from a factorial design are conducted. The set of runs to be conducted is usually selected after considering the confounding which will result.

Fundamental error This is an unavoidable component of sampling variation which arises from the fact that the individual particles vary and most variation between samples is due to the variation between the larger particles in samples which might be selected.

Indicative factor An indicative factor is one which is included in an experiment in order to check that the conclusions of the experiment are valid over the range of that factor. Often such factors will not be controllable in non-experimental situations.

Influential point An influential point is a data item which has a large effect on an important aspect of a fitted model, in the sense that fitting the same type of model but omitting that data item would give a quite different result.

Levels of a factor These are the different values which a factor takes in an experiment.

Interaction An interaction is when the effect of one factor varies with the level of another factor or with the combination of the levels of several other factors. An interaction is sometimes described qualitatively by specifying the effect of that first factor for each of the levels of the second factor (and possibly further factors). At other times, interactions are described qualitatively by specifying how much the effect of that first factor varies with the levels of the second factor (and possibly further factors).

In experiments involving many factors we can speak about two-factor interactions, three-factor interactions and so on. A three-factor interaction between factors A, B and C occurs when the way in which the effect of

GLOSSARY

factor A changes with the level of factor B is not the same for all levels of factor C. Interactions between more than two factors are generally very difficult to comprehend, expect possibly as terms in equations for response surfaces. Note that the unqualified term 'interaction' is sometimes used as an abbreviation for 'two-factor interaction'.

Linear model A linear model is one in which the predictions of the model are linear functions of a set of parameters. Such models can include terms such as Cx^α provided that the coefficient C is being estimated, but the index α is not being estimated.

Lot A lot is a defined quantity of material from which a sample or samples need to be taken, possibly for the purpose of determining some aspect of quality or grade.

Main effects and interactions The main effect of a factor describes the way that the level of that factor influences the trend in responses. If interactions are unimportant then such a description is straightforward. However, if interactions are important then the main effect of a factor strictly means the effect of that factor on the experimental results, averaged over the levels of the other factors used in an experiment.

A possible source of confusion when interactions are important is that the meaning of the term main effect can differ. Knowing the main effect of a factor under one set of conditions does not guarantee that you know the main effect under other conditions.

Observational data This is non-experimental data. It was collected without deliberately varying factors thought likely to influence the data. It is also called historical or happenstance data.

Orthogonality In statistical textbooks, the word 'orthogonal' is frequently used to mean 'at right angles' in the familiar geometrical sense or in mathematical extensions of that sense.

The reason for wanting an experiment to be 'orthogonal' is so that the uncertainties in the estimates of parameters of fitted models will be uncorrelated. One orthogonality condition which is adequate for models which are being used to estimate main effects is for each level of each factor to occur equally often with each level of every other factor. If this condition is satisfied then the design is said to be 'balanced' or 'orthogonal'. Experimental designs commonly recommended for fitting response surface models are not generally 'balanced' in this sense, but they are often 'orthogonal' in the sense that sums of cross products of coded levels of factors are zero.

Outlier An outlier is a data point that seems unlikely according to the fitted model. Generally, this means that the data point is a long way away from the expected value.

Precision Precision is the tendency of a large number of experimental determinations to give consistent answers.

Predictor variable A predictor variable is any quantity which is used in a model for predicting a response. Predictor variables are not necessarily controllable.

Pseudorandom numbers These are numbers generated in a deterministic way, such as by a computer program, which can be regarded as being random for most purposes.

Quantitative A factor or variable is described as quantitative if it can take a range of values on a continuous scale.

Randomization The order of conducting experimental runs is frequently selected in a way such that all orders were equally likely in order to improve the chance that the effects of uncontrolled variables will have little influence on the estimates of the effects of the variables being investigated. Similarly, the assignment of raw materials, plots of soil or measurement apparatus to experimental runs may be randomized.

Regression Regression analysis is a way of fitting linear models. Generally, the criterion for fitting is to minimize the sum of squared residuals. Note that the term 'linear models' has a meaning which is broader that many people expect. It includes models consisting of sums of terms of the form $\sum C_i f_i(x)$ where $f_i(x)$ may be non-linear functions, such as x^3, $\sin(2x)$ or $\exp(-1.2x)$, provided that only the multiplicative constants, C_i, are estimated from the data. If the form of the functions $f_i(x)$ or their parameters are to be estimated from the data such a model would not be described as linear.

Replication This is the conducting of more than one experimental run for a single treatment combination. This allows greater precision and helps in the detection of various types of errors.

Merely repeating part of an experimental protocol (for instance, repeating a measurement but not taking a new sample) does not constitute replication of the experimental run, but it might be described as a 'replicate measurement'.

Residual A residual is the difference between a data point and the corresponding predicted value according to a model.

Resolution of a design This concept is applied mainly to two-level fractional factorial designs. An experimental design is said to be of 'resolution III' if main effects would be confounded with two-factor interactions. Main effects can be estimated under the assumption that all interactions are negligible.

An experimental design is said to be of 'resolution IV' if some main effects are confounded with three-factor interactions and some two-factor interactions are confounded with other two-factor interactions. Main effects can be estimated under the assumption that third and higher order interactions are negligible.

An experimental design is said to be of 'resolution V' if main effects and second order interactions can be estimated under the assumption that third and higher order interactions are negligible.

GLOSSARY

Response A response or a response variable is a measurement, a rating or any other outcome from an experiment or questionnaire. Response variables are also sometimes referred to as output variables or dependent variables.

Sample A sample from a lot is a quantity of material smaller than the entire lot which has resulted from a series of operations intended to make the small quantity of material representative of the entire lot so that measurements made on the sample are likely to be similar to the same measurements made on the entire lot. This usually requires that all parts of the lot being sampled must have the same probability of becoming part of the final sample on which measurements are made.

Often, a carefully prepared sample will only be representative for some purposes, not for all purposes. For instance, a sample which has been crushed and dried might be representative for dry chemical analysis but not representative for moisture determination or considering particle size.

Saturated designs A fractional factorial or other design is said to be saturated if so many effects are to be estimated that there are no degrees of freedom left for the purpose of estimating the error variance. A design is said to be nearly saturated if there are very few degrees of freedom left for the purpose of estimating the error variance.

Split plots These are runs which can be regarded as separate replicates for investigation of some factors but should not be regarded as separate replicates for the investigation of other factors. Experimental designs involving split plots are most commonly used in situations where some factors can only be applied to large amounts of material or are very expensive to adjust, but other factors are easy to adjust or can be applied to smaller amounts of material. Such designs can be regarded as consisting of plots between which the levels of the expensive-to-adjust factors differ, with the easy-to-adjust factors differing on subsets of those plots (which are referred to as 'split plots').

Standard deviation The sample standard deviation is denoted by s and is the square root of the sample variance. The population standard deviation is denoted by σ and is the square root of the population variance.

Statistical model A statistical model includes a description of the trend in a response variable as a function of some predictor variables and a description of the likely departures from this trend. There are often parameters in these descriptions which need to be estimated.

Treatment Treatments are the experimental conditions investigated.

Variance The variance of a set of data $X_1, X_2, \ldots, X_n$ is usually denoted by s^2 and is given by

$$s^2 = \frac{\sum_1^n (X_i - \bar{X})^2}{n-1} \quad \text{where} \quad \bar{X} = \frac{\sum_1^n X_i}{n}.$$

Strictly, s^2 is called the sample variance. It is an unbiased estimate of the average squared departure from the true mean which is called the population variance and is usually written using the Greek letter σ (sigma) as σ^2.

Many computer programs and pocket calculators offer the choice of using either n or $n-1$ as the divisor when calculating sample variances. You should always use $n-1$.[1]

White noise This is noise which shows no correlation between successive values.

---

[1] The divisor n is correct for calculating the population variance for discrete populations in which all outcomes are equally likely. For instance, the possible outcomes of the throw of a single die are 1, 2, 3, 4, 5 and 6. These numbers are not data, but an enumeration of possibilities. The population variance of this set of $n = 6$ numbers is

$$\sigma^2 = \frac{\sum_1^n \left(X_i - \bar{X}\right)^2}{n} = \frac{35}{12}.$$

References

Andersen, A.N., Braithwaite, R.W., Cook, G.D., Corbett, L.K., Williams. R.J., Douglas, M.M., Gill, A.M., Setterfield, S.A., Muller, W.J. (1998) Fire research for conservation management in tropical savannas: Introducing the Kapalga fire experiment. *Australian Journal of Ecology*, **23**, 95–110.

Anscombe, F.J. (1973) Graphs in Statistical Analysis. *The American Statistician*, February 1973.

Barton, R.R. (1997) Pre-experiment planning for designed experiments: Graphical methods. *Journal of Quality Technology*, **29**, 307–316.

Bisgaard, S. and Ankenman, B. (1996) Standard errors for the eigenvalues in second-order response surface models. *Technometrics*, **38**, 238–246.

Bland, J.M. and Altman, D.G. (1986) Statistical methods for assessing agreement between two methods of clinical measurement. *The Lancet*, February 8, 307–310.

Box, G.E.P. (1976) Science and statistics. *Journal of the American Statistical Association*, **71**, 791–799.

Box, G.E.P. (1985) Discussion. *Journal of Quality Technology*, **17**, 189–190.

Box, G.E.P. and Behnken, D.W. (1960) Some new three level designs for the study of quantitative variables. *Technometrics*, **2**, 455–475.

Box, G.E.P., Bisgaard, S. and Fung, C. (1988) An explanation and critique of Taguchi's contributions to quality engineering. *Quality and Reliability Engineering International*, **4**, 123–131.

Box, G.E.P. and Draper, N.R. (1969) *Evolutionary Operation*, Wiley.

Box, G.E.P. and Draper, N.R. (1987) *Empirical Model Building and Response Surfaces*, Wiley.

Box, G.E.P., Hunter, W.G. and Hunter, J.S. (1978) *Statistics for Experimenters*, Wiley.

Box, G.E.P. and Wilson, K.B. (1951) On the experimental attainment of optimum conditions (with discussion). *Journal of the Royal Statistical Society, Series B*, **13**, pages 1–45.

Bryan-Jones, J. and Finney, D.J. (1983) On an error in 'Instructions to Authors'. *HortScience*, **18**, 279–282.

Bussell, W.T., Maindonald, J.H and Morton, J.R. (1997) What is a correct plant density for transplanted green asparagus? *New Zealand Journal of Crop and Horticultural Science*, **25**, 359–368.

Carmer, S.G. and Walker, W.M. (1982) Baby bear's dilemma: A statistical tale. *Agronomy Journal*, **74**, 122–124.

Carter, N.B., Clark, K.N., Thomas, C.G. and Bowling, K.McG. (1983) *Ironmaking and Steelmaking*, **10**, No. 6, 243–252.

Cellier, K.M. and Stace, H.C.T. (1968) The determination of optimum operating conditions in atomic absorption spectroscopy. *Applied Spectroscopy*, **20**, 26–33.

Chambers, J.M. (1993) Greater or lesser statistics: A choice for future research. *StatComp*, **3**, 182–184.

Chan, H.T., Maindonald, J.H., Laidlaw, W.G. and Seltenrich, M. (1996) ACC oxidase in papaya sections after heat treatment. *Journal of Food Science*, **61**(6), 1182–1185.

Cleveland, W.S. (1985) *The Elements of Graphing Data*, Wadsworth.

Cochran, W.C. and Cox, G.M. (1957) *Experimental Designs*, Wiley.

Coleman, D.E. and Montgomery, D.C. (1993) A systematic approach to planning for a designed industrial experiment (with discussion). *Technometrics*, **35**, 1–27.

Cornell, J.A. (1981) *Experiments with Mixtures: Designs, Models and the Analysis of Mixture Data*, Wiley, New York.

Cox, D.R. (1958) *Planning of Experiments*, Wiley.

Creffield, J. and Ngugen, N-K. (1998) *Data analysis on the effectiveness of Silafluoen to protect plywood against attack by subterranean termite*, CSIRO Mathematical and Information Sciences Report Number CMIS 98/83.

Czitrom, V. (1999) One-factor-at-a-time versus designed experiments. *The American Statistician*, **53**, 126–131.

Daniel, C. (1973) One-at-a-time plans. *Journal of the American Statistical Association*, **68**, 353–360.

Daniel, C. (1976) *Applications of Statistics to Industrial Experimentation*, Wiley.

Darby, S.C., Ewart, D.W., Giangrande, P.L.F., Dolin, P.J., Spooner, R.J.D., Rizza, C.R. (1995) Mortality before and after HIV infection in the complete UK population of haemophiliacs. *Nature*, **377**, 79–82.

Davies, O.L. (Ed.) (1967) *The Design and Analysis of Industrial Experiments*, Oliver and Boyd.

Dickinson, E.J. and Robinson, G.K. (1977) *The inter-laboratory precision of the penetration test for paving bitumens when the test is done at $15°C$ with a 200g mass and a 60s loading time*, Australian Road Research Board Research Report ARR No. 73.

Dickinson, T. and Robinson, G.K. (1994) Remember that what you measure affects behaviour. *The Quality Magazine*, February 1994, 83–84.

Doehlert, D.H. (1970) Uniform shell designs. *Applied Statistics*, **19**, 231–239.

Doehlert, D.H. and Klee, V.L. (1972) Experimental designs through level reduction of the d-dimensional cuboctahedron. *Discrete Mathematics*, **2**, 309–334.

Draper, N.R. and Guttman, I. (1980) Incorporating overlap effects from neighbouring units into response surface models. *Applied Statistics*, **29**, 128–134.

Ehrenberg, A.S.C. (1977) Rudiments of numeracy. *Journal of the Royal Statistical Society, Series A*, **140**, 277–297.

Finney, D.J. (1960) *An Introduction to the Theory of Experimental Design*, University of Chicago Press.

Fisher, R.A. (1926) The arrangement of field experiments. *Journal of the Ministry of Agriculture of Great Britain*, **33**, 503–513. Reprinted in *Collected Papers of R.A. Fisher*. Volume II. Edited by J.H. Bennett (1972) The University of Adelaide.

Fisher, R.A. and Yates, F. (1938) *Statistical Tables for Biological, Agricultural and Medical Research*. Oliver and Boyd.

Freedman, D (1999) From association to causation: some remarks on the history of statistics. *Statistical Science*, **14**, 243–258.

Goh, T.N. (1993) Taguchi methods: Some technical, cultural and pedagogical perspectives. *Quality and Reliability Engineering International*, **9**, 185–202.

Gore, S.M. and Altman, D.G. (1982) *Statistics in Practice: Articles published in the British Medical Journal*. British Medical Association, London.

Gould, S.J. (1996) *The Mismeasure of Man* (Revised and expanded.), W.W. Norton & Company Inc, New York.

Gy, P. (1979) *Sampling of Particulate Materials*, Elsevier, Amsterdam.

Hahn, G.J. (1984) Experimental design in the complex world. *Technometrics*, **26**, 19–31.

Hald, A. (1952) *Statistical Tables and Formulas*, Wiley.

Halmos, P.R. (1970) How to write mathematics. *L'Enseignement mathématique*, **XVI**, fasc. 2, 123–152.

Henderson, C.R., Kempthorne, O., Searle, S.R. and von Krosigk, C.N. (1959) Estimation of environmental and genetic trends from records subject to culling. *Biometrics*, **13**, 192–218.

REFERENCES

Hind, P.R., Tritt, B.H. and Hoffman, E.R. (1976) Effects of level of illumination, strokewidth, visual angle and contrast on the legibility of numerals of various fonts. *Australian Road Research Board Proceedings*, **8**, 46–55.

Hunter, J.S. (1988) Design and analysis of experiments. Section 26 of *Juran's Quality Control Handbook*, 4th edn, (eds J.M. Juran and F.M. Gryna), McGraw-Hill.

Hunter, W.G. (1977) Some ideas about teaching design of experiments, with 2^5 examples of experiments conducted by students. *The American Statistician*, **31**, 12–17.

Hunter, W.G. (1981) Six statistical tales. *The Statistician*, **30**, 107–117.

Hunter, W.G. and Kittrell, J.R. (1966) Evolutionary operation: a review. *Technometrics*, **8**, 389–397.

Jeffers, J.N.R. (1978) *Statistical Checklist 1: Design of Experiments*, Natural Environmental Research Council, Institute of Terrestrial Ecology, 68 Hills Road, Cambridge, UK.

Kackar, R.N. (1985) Off-line quality control, Parameter Design and the Taguchi Method. *Journal of Quality Technology*, **17**, 176–209.

Knowlton, J. and Keppinger, R. (1993) The experimentation process. *Quality Progress*, February 1993, 43–47.

Kuhn, T.S. (1970) *The Structure of Scientific Revolutions*, 2nd edn, The University of Chicago Press.

Lin, K.M. and Kackar, R.N. (1986) Optimizing the wave soldering process. *Electronic Packaging & Production*, February, 108–115.

Lucas, J.M. (1976) Which response surface design is best? A performance comparison of several types of quadratic response surface designs in symmetric regions. *Technometrics*, **18**, 411–417.

Maindonald, J.H. and Finch, J.R. (1986) Apple damage from transport in wooden bins. *New Zealand Journal of Technology*, **2**, 171–177.

Mangel, M. and Samaniego, F.J. (1984) Abraham Wald's work on aircraft survivability. *Journal of the American Statistical Association*, **79**, 259–267.

Margolin, B.H. (1969) Results on factorial designs of resolution IV for the 2^n and $2^n 3^m$ series. *Technometrics*, **11**, 431.

Mead, R. (1988) *The Design of Experiments: Statistical Principles for Practical Applications*, Cambridge University Press, Cambridge.

Meyer, D. and Napier-Munn, T. (1999) Optimal experiments for time-dependent mineral processes. *Australian and New Zealand Journal of Statistics*, **41**, 3–17.

Miller, A.J. (1990) *Subset Selection in Regression*, Chapman and Hall.

Pearce, S.C. (1976) *Field Experimentation with Fruit Trees and Other Perennial Plants*, 2nd ed, Commonwealth Agricultural Bureaux.

Philipson, T. and Desimone, J. (1997) Experiments and subject sampling. *Biometrika*, **83**, 619–630.

Pignatiello, J.J. and Ramberg, J.S. (1985) Discussion. *J. Quality Technology*, **17**, 198–206.

Pitard, F.F. (1989) *Pierre Gy's Sampling Theory and Sampling Practice. Volume 1: Heterogeneity and Sampling; Volume 2: Sampling Correctness and Sampling Practice*, CRC Press.

Plackett, R.L. and Burman, J.P. (1946) The design of optimum multifactorial experiments. *Biometrika*, **33**, 305–325.

Polya, G. (1945) *How to Solve It*, Princeton University Press, Princeton.

Rao, C.R. (1997) *Statistics and Truth: Putting Chance to Work*, 2nd edn, World Scientific Publishing, Singapore.

Robinson, G.K. (1993a) Box–Cox transformations and Taguchi Method, (Letter to the Editors). *Applied Statistics*, **42**, 557–558.

Robinson, G.K. and Holmes, R.J. (1995) *Iron ores – Results of testwork on sample division for iron ores. ISO/TC 102 – Iron ores*, International Standards Organization, TC 102 Technical Committee Report No. 9.

Russell, W.M.S. and Burch, R.L. (1959) *The Principles of Humane Experimental Technique*, Methuen, London.

Salsburg, D.S. (1973) Sufficiency and the waste of information. *The American Statistician*, **27**, 152–154.

Saunders, I.W. and Eccleston, J.A. (1992) Experimental design for continuous processes. *Australian Journal of Statistics*, **34**, 77–89.

Scheaffer, R.L. (1997) Discovery of sampling concepts through activities. *Bulletin of the International Statistical Institute. 51st Session.* Proceedings Tome **LVII**, Book 1, 421–424.

Senn, S. (1998) Mathematics: governess or handmaiden? *The Statistician*, **47**, 241–259.

Shainin, D. and Shainin, P. (1988) Better than Taguchi Orthogonal Tables. *Quality and Reliability Engineering International*, **4**, 143–149.

Snee, R.D., Hare, L.B. and Trout, J.R. (1985) *Experiments in Industry – Design, Analysis and Interpretation of Results*, American Society for Quality Control.

Steinberg, D.M. (1988) Factorial experiments with time trends. *Technometrics*, **30**, 259–269.

'Student' (1931) The Lanarkshire milk experiment. *Biometrika*, **23**, 398–406.

Taguchi, G. (1986) *Introduction to Quality Engineering: Designing Quality into Products and Processes*, Asian Productivity Association.

Taguchi, G. (1987) *System of Experimental Design: Engineering Methods to Optimize Quality and Minimize Costs*, two volumes, UNIPUB/Kraus International Publications.

Tanur, J.M., Mosteller, F., Kruskal, W.H., Link, R.F., Pieters, R.S. and Rising, G.R. (1972) *Statistics: A Guide to the Unknown*, Holden-Day.

Todhunter, I. (1949) *A History of the Mathematical Theory of Probability from the Time of Pascal to that of Laplace*, Chelsea Pub. Co., New York.

Tufte, E.R. (1983) *The Visual Display of Quantitative Information*, Graphics Press, Cheshire, Connecticut.

Tukey, J.W. (1962) The future of data analysis. *Annals of Mathematical Statistics*, **33**, 1–67.

Verbyla, A.P., Cullis, B.R., Kenward, M.G. and Welham, S.J. (1999) The analysis of designed experiments and longitudinal data by using smoothing splines (with discussion). *Applied Statistics*, **48**, 269–311.

Wainer, H. (1984) How to display data badly. *The American Statistician*, **38**, 137–147.

Wallis, W.A. and Roberts, H.V. (1956) *Statistics: A New Approach*, Free Press, Glencoe, Illinois.

Whitney, J.B. and Young, J.C. (1989) Statistical methodology: prerequisites for effective implementation in industry. *Philosophical Transactions of the Royal Society of London*, **A 327**, 591–598.

Index

Absorption 41
Accuracy 251
Agricultural trials 24, 62, 124
Analysis of variance 251
Animal trials 24
Applicability
   ensuring general 125
Asking questions 14, 21
Assumptions 219
Automatic process control 19, 44
Average 251

Balance 255
Balanced incomplete block designs 173
Balanced nested designs 181
Biases 43, 78, 127, 251
   in selection 145, 146
Blinds 132
Block 251
Blocking 121, 251
Box–Behnken designs 165

Carryover designs 173
Catalogues of designs 179
Causation 30
Central composite designs 165
Centre points 88
Checking data is reasonable 200
Complexity of models 68
Components of noise 239
Components of variation 24, 118, 149
Confounding 143, 157, 252
Context 15
Continuous variables 252
Contrasts 212, 252
Controls 12, 177
Correlation 31, 232, 252
Costs of experimenting 61, 124, 152
Covariates 121, 252
Customers 48, 114

Data analysis 199
Data quality 200
Data-recording 189

Degrees of freedom 149, 153, 225, 253
Dependent variable 22, 257
Designs robust against autocorrelated noise 174
Discrete variables 253
Double blind experiments 132

Efficiency 116
Environmental issues 20, 58, 196
Error variance 22, 40, 76, 125, 231
Errors of type I and type II 23
Ethical issues 20, 58, 196
EVOP 88, 91, 92, 253
Examination questions 249
Examples
   aircraft safety and logos 146
   aircraft survivability 44
   apple damage from transport in wooden bins 240
   atomic absorption spectrometers 67
   bank robberies 19, 31
   basic oxygen steelmaking furnace 33
   bias of a flap sampler 122
   blending of powders 37
   cake, making a 57
   checking bias and precision of iron ore sample preparation and testing 78
   coffees and whiskies 68
   comparing gold assaying procedures 231
   comparing logit and probit models 228
   density of asparagus plants 39
   disassembly and reassembly 159
   dissolution of bauxite 193, 214
   election results 119
   estimating parameters for modelling 15
   fat depth 232
   fire ecology experiment 14
   fruit tree spraying 117
   genetic merit of dairy cattle 201
   gentamicin 93
   gold grades 229
   gold recovery 136
   grapes 12, 21

health and cigarette consumption 32
heat treatment of papayas 118
HIV and haemophilia 245
hydrocyclones 56
Ina tile experiments 181
interpretation of road signs 120
Lanarkshire milk experiment 135
leaf springs for trucks 26
mixing paint 57
mode choice after Tasman bridge accident 225
okra variety trial 220
orientations of planetary orbits 230
photography 54
plasma etching 57
preservation of grapes 12, 21
quality of cement 118
raspberry canes 18
rats losing weight 139
relays for telephone switchboards 67
road maintenance strategies 29
salk polio vaccine 133
scoring decay in timber 49
SCRIM comparison 63
simulated data 211, 216, 233
skid resistance of tiles 49
soil with sugar cane 131
staff appointments 31
steering strategies 246
sugar cane sampling error 76
sulphur dioxide fumigation 12, 21
survey bias 43
tensile testing of welds in plastic pipe 224
termites attacking plywood 177
testing bitumen 76, 216
testing insecticides 178
visibility of digits 203
washing powder advertisements 143
wave soldering machine 59, 81
weeds in plots 133
weighing shipments of iron ore 131
Experimental error 22, 75, 253
Experimental protocols 58, 62, 199
Experimental runs 148, 253
Experimental strategy 61, 239
Extraction ratio 130

Factor 253
Factor screening 23, 50, 65, 154, 175
Factorial designs 155, 254
Factors 50, 175
  with three levels 161, 208
  with two levels 156, 207

Flow charts 15
Fit, goodness of 233
Foldover designs 159, 254
Formal statistical analysis 200, 242
Fractional factorial designs 156, 161, 254
Fundamental error 35, 254

Goal Hierarchy Plots 15
Graphs 200, 221, 244
Greengrocer test 44

Half-baked ideas 10
Happenstance data 255
Hierarchical experiments 77, 181
Historical data 31, 39, 255
  cautions 41
HIV and haemophilia 245

Independence 31, 227
Independent variable 22, 254
Indicative factors 175, 254
Influence matrix 58
Influential point(s) 220, 254
Interaction(s) 54, 69, 206, 255
Interlaboratory trials 78, 181
Intermediates 49

Labelling of samples 190
Laboratory and plant trials 17, 124, 126
Latin squares 172
Least squares 236
Levels of a factor 53, 175, 254
Linear model 255
Linear terms 69
Lot 255

Main effect 67, 255
Main effects experiments 23, 79, 175, 180
Managers of experimenters 248
'Maybes' 10
Medical trials 23, 24, 62
Missing data 42, 201
Multiple comparisons 212

Neighbour designs 174
Number of runs 117, 148

Objective 9, 23, 239
Observational data 41, 255
Observational studies 31
Off-line experimentation 87
One-factor-at-a-time experimentation 71
On-line experimentation 87
Operational definition 51

INDEX

Orthogonal 144, 255
Orthogonal arrays 156, 161
Outlier(s) 200, 220, 255
Output variable 257

Paradigm change 4
Parameter design 25
Partial correlation 32, 252
Precision 75, 255
Predictor variable 256
Problem owner 11, 248
Problem-solving model 3
Process view 19, 31, 48
Protocols 58, 199
Proxy measures 51
Pseudorandom numbers 136, 256

Quadratic terms 69, 89
Quadratic loss function 26, 152
Qualitative variable(s) 254
Quantitative variable(s) 254, 256
Questioning 14, 39, 50

Radiation counts 34
Randomization 120, 134, 136, 256
Range of application 18, 125
Range of validity 67
Recording results 189, 196
Reducing error variance 118, 125
Regression 88, 211, 220, 256
Reparametrizing factors 54
Replication 119, 148, 256
Representative sample 35
Reproducibility 125, 231
Residual(s) 225, 236, 256
Resolution of a design 157, 256
Response 48, 257
Response surface experiment(s) 24, 87, 90, 165, 177
Response surfaces
    fitting 88, 205
Response variables 22, 48, 257
Runs 148, 253

Safety 20, 56, 58, 196
Sample 35, 257
Sampling 34, 127
Sampling and testing variation 23, 75, 175, 181
Saturated designs 225, 257
Scaling up 126
Scientific method 4, 71
Selected data 42

Sequential experimentation 17, 61, 65, 88, 239
Signal 122
Signal-to-noise ratio 25, 116
Sizing 38
Slurry sampling 129
Soft and hard science 22
Split plots 123, 235, 257
Spreadsheets 191
Standard deviation 257
Standardization 133
Star points 88
Statistical analysis 204
Statistical models 22, 257
Stepwise regression 235
Strategy 61, 239
Subjective measurements 49, 116
Supplies 194

Tables 243
Taguchi 25
    orthogonal arrays 161
    parameter design 25
    tolerance design 25
Terminology 15, 35, 67, 251
Theoretical models 24, 32
Tolerance design 25
Training 2
Transformations 51
Treatment 175, 257
Types of data 50
Types of experiments 23

Using computers for statistical analysis 234

Variance 257

What materials to use? 124
What to measure? 48, 113, 232
When to experiment? 124
Where to experiment? 124
White noise 258

WILEY SERIES IN PROBABILITY AND STATISTICS
ESTABLISHED BY WALTER A. SHEWHART AND SAMUEL S. WILKS

Editors
*Vic Barnett, Noel A. C. Cressie, Nicholas I. Fisher, Iain M. Johnstone,
J. B. Kadane, David W. Scott, Bernard W. Silverman,
Adrian F. M. Smith, Jozef L. Teugels, Ralph A. Bradley, Emeritus,
J. Stuart Hunter, Emeritus, David G. Kendall, Emeritus*

Probability and Statistics Section

*ANDERSON · The Statistical Analysis of Time Series
ARNOLD, BALAKRISHNAN, and NAGARAJA · A First Course in Order Statistics
ARNOLD, BALAKRISHNAN, and NAGARAJA · Records
BACCELLI, COHEN, OLSDER, and QUADRAT · Synchronization and Linearity: An Algebra for Discrete Event Systems
BARNETT · Comparative Statistical Inference, *Third Edition*
BASILEVSKY · Statistical Factor Analysis and Related Methods: Theory and Applications
BERNARDO and SMITH · Bayesian Theory
BILLINGSLEY · Convergence of Probability Measures
BOROVKOV · Asymptotic Methods in Queuing Theory
BOROVKOV · Ergodicity and Stability of Stochastic Processes
BRANDT, FRANKEN, and LISEK · Stationary Stochastic Models
CAINES · Linear Stochastic Systems
CAIROLI and DALANG · Sequential Stochastic Optimization
CONSTANTINE · Combinatorial Theory and Statistical Design
COOK · Regression Graphics
COVER and THOMAS · Elements of Information Theory
CSÖRGŐ and HORVÁTH · Weighted Approximation in Probability Statistics
CSÖRGŐ and HORVÁTH · Limit Theorems in Change Point Analysis
DETTE and STUDDEN · The Theory of Canonical Moments with Applications in Statistics, Probability, and Analysis
*DOOB · Stochastic Processes
DUPUIS and ELLIS · A Weak Convergence Approach to the Theory of Large Deviations
ETHIER and KURTZ · Markov Processes: Characterization and Convergence
FELLER · An Introduction to Probability Theory and Its Applications, Volume 1, *Third Edition*, Revised; Volume II, *Second Edition*
FULLER · Introduction to Statistical Time Series, *Second Edition*
FULLER · Measurement Error Models
GHOSH, MUKHOPADHYAY, and SEN · Sequential Estimation
GIFI · Nonlinear Multivariate Analysis
GUTTORP · Statistical Inference for Branching Processes
HALL · Introduction to the Theory of Coverage Processes

---

*Now available in a lower priced paperback edition in the Wiley Classics Library.

Probability and Statistics Section (Continued)
HAMPEL · Robust Statistics: The Approach Based on Influence Functions
HANNAN and DEISTLER · The Statistical Theory of Linear Systems
HUBER · Robust Statistics
HUŠKOVÁ, BERAN, and DUPAČ · Collected Works of Jaroslav Hájek—With Commentary
IMAN and CONOVER · A Modern Approach to Statistics
JUREK and MASON · Operator-Limit Distributions in Probability Theory
KASS and VOS · The Geometrical Foundations of Asymptotic Inference
KAUFMAN and ROUSSEEUW · Finding Groups in Data: An Introduction to Cluster Analysis
KELLY · Probability, Statistics, and Optimization
KENDALL, BARDEN, CARNE, and LE · Shape and Shape Theory
LINDVALL · Lectures on the Coupling Method
McFADDEN · Management of Data in Clinical Trials
MANTON, WOODBURY, and TOLLEY · Statistical Applications Using Fuzzy Sets
MARDIA and JUPP · Directional Statistics
MORGENTHALER and TUKEY · Configural Polysampling: A Route to Practical Robustness
MUIRHEAD · Aspects of Multivariate Statistical Theory
OLIVER and SMITH · Influence Diagrams, Belief Nets, and Decision Analysis
*PARZEN · Modern Probability Theory and Its Applications
PRESS · Bayesian Statistics: Principles, Models, and Applications
PUKELSHEIM · Optimal Experimental Design
RAO · Asymptotic Theory of Statistical Inference
RAO · Linear Statistical Inference and Its Applications, *Second Edition*
RAO and SHANBHAG · Choquet-Deny Type Functional Equations with Applications to Stochastic Models
ROBERTSON, WRIGHT, and DYKSTRA · Order Restricted Statistical Inference
ROGERS and WILLIAMS · Diffusions, Markov Processes, and Martingales, Volume I: Foundations, *Second Edition*, Volume II: Itô Calculus
RUBINSTEIN and SHAPIRO · Discrete Event Systems: Sensitivity Analysis and Stochastic Optimization by the Score Function Method
RUZSA and SZEKELEY · Algebraic Probability Theory
SCHEFFÉ · The Analysis of Variance
SEBER · Linear Regression Analysis
SEBER · Multivariate Observations
SEBER and WILD · Nonlinear Regression
SERFLING · Approximation Theorems of Mathematical Statistics
SHORACK and WELLNER · Empirical Processes with Applications to Statistics
SMALL and McLEISH · Hilbert Space Methods in Probability and Statistical Inference
STAPLETON · Linear Statistical Models
STAUDTE and SHEATHER · Robust Estimation and Testing
STOYANOV · Counterexamples in Probability
TANAKA · Time Series Analysis: Nonstationary and Noninvertible Distribution Theory
THOMPSON and SEBER · Adaptive Sampling
WELSH · Aspects of Statistical Inference
WHITTAKER · Graphical Models in Applied Multivariate Statistics
YANG · The Construction Theory of Denumerable Markov Processes

---

*Now available in a lower priced paperback edition in the Wiley Classics Library.

Applied Probability and Statistics Section

ABRAHAM and LEDOLTER · Statistical Methods for Forecasting
AGRESTI · Analysis of Ordinal Categorical Data
AGRESTI · Categorical Data Analysis
ANDERSON, AUQUIER, HAUCH, OAKES, VANDAELE, and WEISBERG · Statistical Methods for Comparative Studies
ARMITAGE and DAVID (editors) · Advances in Biometry
*ARTHANARI and DODGE · Mathematical Programming in Statistics
ASMUSSEN · Applied Probability and Queues
*BAILEY · The Elements of Stochastic Processes with Applications to the Natural Sciences
BARNETT and LEWIS · Outliers in Statistical Data, *Third Edition*
BARTHOLOMEW, FORBES, and McLEAN · Statistical Techniques for Manpower Planning, *Second Edition*
BATES and WATTS · Nonlinear Regression Analysis and Its Applications
BECHHOFER, SANTNER, and GOLDSMAN · Design and Analysis of Experiments for Statistical Selection, Screening, and Multiple Comparisons
BELSLEY · Conditioning Diagnostics: Collinearity and Weak Data in Regression
BELSLEY, KUH, and WELSCH · Regression Diagnostics: Identifying Influential Data and Sources of Collinearity
BHAT · Elements of Applied Stochastic Processes, *Second Edition*
BHATTACHARYA and WAYMIRE · Stochastic Processes with Applications
BIRKES and DODGE · Alternative Methods of Regression
BLISCHKE and PRABHAKAR MURTHY · Reliability: Modeling, Prediction, and Optimization
BLOOMFIELD · Fourier Analysis of Time Series: An Introduction
BOLLEN · Structural Equations with Latent Variables
BOULEAU · Numerical Methods for Stochastic Processes
BOX · Bayesian Inference in Statistical Analysis
BOX and DRAPER · Empirical Model-Building and Response Surfaces
BOX and DRAPER · Evolutionary Operation: A Statistical Method for Process Improvement
BUCKLEW · Large Deviation Techniques in Decision, Simulation, and Estimation
BUNKE and BUNKE · Nonlinear Regression, Functional Relations, and Robust Methods: Statistical Methods of Model Building
CHATTERJEE and HADI · Sensitivity Analysis in Linear Regression
CHERNICK · Bootstrap Methods: A Practitioner's Guide
CHILÈS and DELFINER · Geostatistics: Modeling Spatial Uncertainty
CHOW and LIU · Design and Analysis of Clinical Trials
CLARKE and DISNEY · Probability and Random Processes: A First Course with Applications, *Second Edition*
*COCHRAN and COX · Experimental Designs, *Second Edition*
CONOVER · Practical Nonparametric Statistics, *Second Edition*
CORNELL · Experiments with Mixtures, Designs, Models, and the Analysis of Mixture Data, *Second Edition*
*COX · Planning of Experiments
CRESSIE · Statistics for Spatial Data, *Revised Edition*
DANIEL · Applications of Statistics to Industrial Experimentation
DANIEL · Biostatistics: A Foundation for Analysis in the Health Sciences, *Sixth Edition*
DAVID · Order Statistics, *Second Edition*
*DEGROOT, FIENBERG, and KADANE · Statistics and the Law
*DODGE · Alternative Methods of Regression

---

*Now available in a lower priced paperback edition in the Wiley Classics Library.

Applied Probability and Statistics (Continued)
  DOWDY and WEARDEN · Statistics for Research, *Second Edition*
  DRYDEN and MARDIA · Statistical Shape Analysis
  DUNN and CLARK · Applied Statistics: Analysis of Variance and Regression, *Second Edition*
  ELANDT-JOHNSON and JOHNSON · Survival Models and Data Analysis
  EVANS, PEACOCK, and HASTINGS · Statistical Distributions, *Second Edition*
  FLEISS · The Design and Analysis of Clinical Experiments
  FLEISS · Statistical Methods for Rates and Proportions, *Second Edition*
  FLEMING and HARRINGTON · Counting Processes and Survival Analysis
  GALLANT · Nonlinear Statistical Models
  GLASSERMAN and YAO · Monotone Structure in Discrete-Event Systems
  GNANADESIKAN · Methods for Statistical Data Analysis of Multivariate Observations. *Second Edition*
  GOLDSTEIN and LEWIS · Assessment: Problems, Development, and Statistical Issues
  GREENWOOD and NIKULIN · A Guide to Chi-squared Testing
 *HAHN · Statistical Models in Engineering
  HAHN and MEEKER · Statistical Intervals: A Guide for Practitioners
  HAND · Construction and Assessment of Classification Rules
  HAND · Discrimination and Classification
  HEDAYAT and SINHA · Design and Inference in Finite Population Sampling
  HEIBERGER · Computation for the Analysis of Designed Experiments
  HINKELMAN and KEMPTHORNE · Design and Analysis of Experiments, Volume 1: Introduction to Experimental Design
  HOAGLIN, MOSTELLER, and TUKEY · Exploratory Approach to Analysis of Variance
  HOAGLIN, MOSTELLER, and TUKEY · Exploring Data Tables, Trends and Shapes
  HOAGLIN, MOSTELLER, and TUKEY · Understanding Robust and Exploratory Data Analysis
  HOCHBERG and TAMHANE · Multiple Comparison Procedures
  HOCKING · Methods and Applications of Linear Models: Regression and the Analysis of Variables
  HOGG and KLUGMAN · Loss Distributions
  HOLLANDER and WOLFE · Nonparametric Statistical Methods
  HOSMER and LEMESHOW · Applied Logistic Regression
  HØYLAND and RAUSAND · System Reliability Theory: Models and Statistical Methods
  HUBERTY · Applied Discriminant Analysis
  HUNT and KENNEDY · Financial Derivatives in Theory and Practice
  JACKSON · A User's Guide to Principal Components
  JOHN · Statistical Methods in Engineering and Quality Assurance
  JOHNSON · Multivariate Statistical Simulation
  JOHNSON & KOTZ · Distributions in Statistics
           Continuous Multivariate Distributions
  JOHNSON, KOTZ, and BALAKRISHNAN · Continuous Univariate Distributions, Volume 1, *Second Edition*
  JOHNSON, KOTZ, and BALAKRISHNAN · Continuous Univariate Distributions, Volume 2, *Second Edition*
  JOHNSON, KOTZ, and BALAKRISHNAN · Discrete Multivariate Distributions
  JOHNSON, KOTZ, and KEMP · Univariate Discrete Distributions, *Second Edition*
  JUREČKOVÁ and SEN · Robust Statistical Procedures: Asymptotics and Interrelations
  KADANE · Bayesian Methods and Ethics in a Clinical Trial Design
  KADANE and SCHUM · A Probabilistic Analysis of the Sacco and Vanzetti Evidence
  KALBFLEISCH and PRENTICE · The Statistical Analysis of Failure Time Data

*Now available in a lower priced paperback edition in the Wiley Classics Library.

Applied Probability and Statistics (Continued)

KELLY · Reversibility and Stochastic Networks
KHURI, MATHEW, and SINHA · Statistical Tests for Mixed Linear Models
KLUGMAN, PANJER, and WILLMOT · Loss Models: From Data to Decisions
KLUGMAN, PANJER, and WILLMOT · Solutions Manual to Accompany Loss Models: From Data to Decisions
KOVALENKO, KUZNETZOV, and PEGG · Mathematical Theory of Reliability of Time-Dependent Systems with Practical Applications
LAD · Operational Subjective Statistical Methods: A Mathematical, Philosophical, and Historical Introduction
LANGE, RYAN, BILLARD, BRILLINGER, CONQUEST, and GREENHOUSE · Case Studies in Biometry
LAWLESS · Statistical Models and Methods for Lifetime Data
LEE · Statistical Methods for Survival Data Analysis, *Second Edition*
LEPAGE and BILLARD · Exploring the Limits of Bootstrap
LINHART and ZUCCHINI · Model Selection
LITTLE and RUBIN · Statistical Analysis with Missing Data
MAGNUS and NEUDECKER · Matrix Differential Calculus with Applications in Statistics and Econometrics, *Revised Edition*
MALLER and ZHOU · Survival Analysis with Long Term Survivors
MANN, SCHAFER, and SINGPURWALLA · Methods for Statistical Analysis of Reliability and Life Data
McLACHLAN and KRISHNAN · The EM Algorithm and Extensions
McLACHLAN · Discriminant Analysis and Statistical Pattern Recognition
McNEIL · Epidemiological Research Methods
MEEKER and ESCOBAR · Statistical Methods for Reliability Data
MILLER · Survival Analysis
MONTGOMERY and PECK · Introduction to Linear Regression Analysis, *Second Edition*
MYERS and MONTGOMERY · Response Surface Methodology: Process and Product in Optimization Using Designed Experiments
NELSON · Accelerated Testing, Statistical Models, Test Plans, and Data Analyses
NELSON · Applied Life Data Analysis
OCHI · Applied Probability and Stochastic Processes in Engineering and Physical Sciences
OKABE, BOOTS, CHUI, and SUGIHARA · Spatial Tessellations, *Second Edition*
PANKRATZ · Forecasting with Dynamic Regression Models
PANKRATZ · Forecasting with Univariate Box–Jenkins Models: Concepts and Cases
PIANTADOSI · Clinical Trials: A Methodologic Perspective
PORT · Theoretical Probability for Applications
PUTERMAN · Markov Decision Processes: Discrete Stochastic Dynamic Programming
RACHEV · Probability Metrics and the Stability of Stochastic Models
RÉNYI · A Diary on Information Theory
RIGDON and BASU · Statistical Methods for the Reliability of Repairable Systems
RIPLEY · Spatial Statistics
RIPLEY · Stochastic Simulation
ROLSKI, SCHMIDLI, SCHMIDT, and TEUGELS · Stochastic Processes for Insurance and Finance
ROUSSEEUW and LEROY · Robust Regression and Outlier Detection
RUBIN · Multiple Imputation for Nonresponse in Surveys
RUBINSTEIN · Simulation and the Monte Carlo Method
RUBINSTEIN and MELAMED · Modern Simulation and Modeling
RYAN · Statistical Methods for Quality Improvement

---

*Now available in a lower priced paperback edition in the Wiley Classics Library.

Applied Probability and Statistics (Continued)

SCHUSS · Theory and Applications of Stochastic Differential Equations
SCOTT · Multivariate Density Estimation: Theory, Practice, and Visualization
*SEARLE · Linear Models
SEARLE · Linear Models for Unbalanced Data
SEARLE, CASELLA, and McCULLOCH · Variance Components
STOYAN, KENDALL, and MECKE · Stochastic Geometry and Its Applications, Second Edition
STOYAN and STOYAN · Fractals, Random Shapes, and Point Fields: Methods of Geometrical Statistics
THOMPSON · Empirical Model Building
THOMPSON · Sampling
THOMPSON · Simulation: A Modeler's Approach
TIJMS · Stochastic Modeling and Analysis: A Computational Approach
TIJMS · Stochastic Models: An Algorithmic Approach
TITTERINGTON, SMITH, and MARKOV · Statistical Analysis of Finite Mixture Distributions
UPTON and FINGLETON · Spatial Data Analysis by Example, Volume 1: Point Pattern and Quantitative Data
UPTON and FINGLETON · Spatial Data Analysis by Example, Volume II: Categorical and Directional Data
VAN RIJKEVORSEL and DE LEEUW · Component and Correspondence Analysis
VIDAKOVIC · Statistical Modeling by Wavelets
WEISBERG · Applied Linear Regression, *Second Edition*
WESTFALL and YOUNG · Resampling-Based Multiple Testing: Examples and Methods for *p*-Value Adjustment
WHITTLE · Systems in Stochastic Equilibrium
WOODING · Planning Pharmaceutical Clinical Trials: Basic Statistical Principles
WOOLSON · Statistical Methods for the Analysis of Biomedical Data
WU and HAMANDA · Experiments: Planning, Analysis, and Parameter Design Optimization
*ZEELNER · An Introduction to Bayesian Inference in Econometrics

Texts and References Section

AGRESTI · An Introduction to Categorical Data Analysis
ANDERSON · An Introduction to Multivariate Statistical Analysis, *Second Edition*
ANDERSON and LOYNES · The Teaching of Practical Statistics
ARMITAGE and COLTON · Encyclopedia of Biostatistics. 6 Volume set
BARTOSZYNSKI and NIEWIADOMSKA-BUGAJ · Probability and Statistical Inference
BENDAT and PIERSOL · Random Data: Analysis and Measurement Procedures, *Third Edition*
BERRY, CHALONER, and GEWEKE · Bayesian Analysis in Statistics and Econometrics: Essays in Honor of Arnold Zellner
BHATTACHARYA and JOHNSON · Statistical Concepts and Methods
BILLINGSLEY · Probability and Measure, *Second Edition*
BOX · R. A. Fisher, the Life of a Scientist
BOX, HUNTER, and HUNTER · Statistics for Experimenters: An Introduction to Design, Data Analysis, and Model Building
BOX and LUCEÑO · Statistical Control by Monitoring and Feedback Adjustment
BROWN and HOLLANDER · Statistics: A Biomedical Introduction

---

*Now available in a lower priced paperback edition in the Wiley Classics Library.

Texts and References Section (Continued)
CHATTERJEE and PRICE · Regression Analysis by Example, *Second Edition*
COOK and WEISBERG · An Introduction to Regression Graphics
COOK and WEISBERG · Applied Regression Including Computing and Graphics
COX · A Handbook of Introductory Statistical Methods
DILLON and GOLDSTEIN · Multivariate Analysis: Methods and Applications
DODGE and ROMIG · Sampling Inspection Tables, *Second Edition*
DRAPER and SMITH · Applied Regression Analysis, *Third Edition*
DUDEWICZ and MISHRA · Modern Mathematical Statistics
DUNN · Basic Statistics: A Primer for the Biomedical Sciences, *Second Edition*
FISHER and VAN BELLE · Biostatistics: A Methodology for the Health Sciences
FREEMAN and SMITH · Aspects of Uncertainty: A Tribute to D. V. Lindley
GROSS and HARRIS · Fundamentals of Queueing Theory, *Third Edition*
HALD · A History of Probability and Statistics and their Applications Before 1750
HALD · A History of Mathematical Statistics from 1750 to 1930
HELLER · MACSYMA for Statisticians
HOEL · Introduction to Mathematical Statistics, *Fifth Edition*
HOLLANDER and WOLFE · Nonparametric Statistical Methods, *Second Edition*
HOSMER and LEMESHOW · Applied Survival Analysis: Regression Modeling of Time to Event Data
JOHNSON and BALAKRISHNAN · Advances in the Theory and Practice of Statistics: A Volume in Honor of Samuel Kotz
JOHNSON and KOTZ (editors) · Leading Personalities in Statistical Sciences: From the Seventeenth Century to the Present
JUDGE, GRIFFITHS, HILL, LÜTKEPOHL, and LEE · The Theory and Practice of Econometrics, *Second Edition*
KHURI · Advanced Calculus with Applications in Statistics
KOTZ and JOHNSON (editors) · Encyclopedia of Statistical Sciences. Volumes 1 to 9 with Index
KOTZ and JOHNSON (editors) · Encyclopedia of Statistical Sciences: Supplement Volume
KOTZ, REED, and BANKS (editors) · Encyclopedia of Statistical Sciences: Update Volume 1
KOTZ, REED, and BANKS (editors) · Encyclopedia of Statistical Sciences: Update Volume 2
LAMPERTI · Probability: A Survey of the Mathematical Theory, *Second Edition*
LARSON · Introduction to Probability Theory and Statistical Inference, *Third Edition*
LE · Applied Categorical Data Analysis
LE · Applied Survival Analysis
MALLOWS · Design, Data, and Analysis by Some Friends of Cuthbert Daniel
MARDIA · The Art of Statistical Science: A Tribute to G. S. Watson
MASON, GUNST, and HESS · Statistical Design and Analysis of Experiments with Applications to Engineering and Science
MURRAY · X-STAT 2.0 Statistical Experimentation, Design Data Analysis, and Nonlinear Optimization
PURI, VILAPLANA, and WERTZ · New Perspectives in Theoretical and Applied Statistics
RENCHER · Methods of Multivariate Analysis
RENCHER · Multivariate Statistical Inference with Applications
ROBINSON · Practical Strategies for Experimenting
ROSS · Introduction to Probability and Statistics for Engineers and Scientists
ROHATGI · An Introduction to Probability Theory and Mathematical Statistics
RYAN · Modern Regression Methods

*Now available in a lower priced paperback edition in the Wiley Classics Library.

Texts and References Section (Continued)
SCHOTT · Matrix Analysis for Statistics
SEARLE · Matrix Algebra Useful for Statistics
STYAN · The Collected Papers of T. W. Anderson: 1943–1985
TIERNEY · LISP-STAT: An Object-Oriented Environment for Statistical Computing and Dynamic Graphics
WONNACOTT and WONNACOTT · Econometrics, *Second Edition*

WILEY SERIES IN PROBABILITY AND STATISTICS
ESTABLISHED BY WALTER A. SHEWHART AND SAMUEL S. WILKS

Editors
Robert M. Groves, Graham Kalton, J. N. K. Rao, Norbert Schwarz, Christopher Skinner

Survey Methodology Section

BIEMER, GROVES, LYBERG, MATHIOWETZ, and SUDMAN · Measurement Errors in Surveys
COCHRAN · Sampling Techniques, *Third Edition*
COUPER, BAKER, BETHLEHEM, CLARK, MARTIN, NICHOLLS and O'REILLY (editors) · Computer Assisted Survey Information Collection
COX, BINDER, CHINNAPPA, CHRISTIANSON, COLLEDGE, and KOTT (editors) · Business Survey Methods
*DEMING · Sample Design in Business Research
DILLMAN · Mail and Telephone Surveys: The Total Design Method
GROVES · Survey Errors and Survey Costs
GROVES and COUPER · Nonresponse in Household Interview Surveys
GROVES, BIEMER, LYBERG, MASSEY, NICHOLLS, and WAKSBERG · Telephone Survey Methodology
*HANSEN, HURWITZ, and MADOW · Sample Survey Methods and Theory, Volume I: Methods and Applications
*HANSEN, HURWITZ, and MADOW · Sample Survey Methods and Theory, Volume II: Theory
KASPRZYK, DUNCAN, KALTON, and SINGH · Panel Surveys
KISH · Statistical Design for Research
*KISH · Survey Sampling
LESSLER and KALSBEEK · Nonsampling Error in Surveys
LEVY and LEMESHOW · Sampling of Populations: Methods and Applications
LYBERG, BIEMER, COLLINS, de LEEUW, DIPPO, SCHWARZ, TREWIN (editors) · Survey Measurement and Process Quality
SIRKEN, HERRMANN, SCHECHTER, SCHWARZ, TANUR and TOURNANGEAU (editors) · Cognition and Survey Research
SKINNER, HOLT and SMITH · Analysis of Complex Surveys

---

*Now available in a lower priced paperback edition in the Wiley Classics Library.